Wave and Tidal Generation Devices, 2nd Edition

Other related titles:

You may also like

- PBRN018 | Peter Tavner | Wave and Tidal Generation Devices, First Edition | 2017
- PBPO129 | Domenico P. Coiro and Tonio Sant | Renewable Energy from the Oceans | 2019
- PBRN017 | Raymond Alcorn and Dara O'Sullivan | Electrical Design for Ocean Wave and Tidal Energy Systems | 2013
- PBPO201 | Mohamed Benbouzid | Generator Systems for Tidal Power: Design, control, and monitoring | 2023

We also publish a wide range of books on the following topics:
Computing and Networks
Control, Robotics and Sensors
Electrical Regulations
Electromagnetics and Radar
Energy Engineering
Healthcare Technologies
History and Management of Technology
IET Codes and Guidance
Materials, Circuits and Devices
Model Forms
Nanomaterials and Nanotechnologies
Optics, Photonics and Lasers
Production, Design and Manufacturing
Security
Telecommunications
Transportation

All books are available in print via https://shop.theiet.org or as eBooks via our Digital Library https://digital-library.theiet.org.

IET ENERGY ENGINEERING SERIES 249

Wave and Tidal Generation Devices, 2nd Edition

Reliability and availability

Peter Tavner

The Institution of Engineering and Technology

About the IET

This book is published by the Institution of Engineering and Technology (The IET).

We inspire, inform and influence the global engineering community to engineer a better world. As a diverse home across engineering and technology, we share knowledge that helps make better sense of the world, to accelerate innovation and solve the global challenges that matter.

The IET is a not-for-profit organisation. The surplus we make from our books is used to support activities and products for the engineering community and promote the positive role of science, engineering and technology in the world. This includes education resources and outreach, scholarships and awards, events and courses, publications, professional development and mentoring, and advocacy to governments.

To discover more about the IET please visit https://www.theiet.org/

About IET books

The IET publishes books across many engineering and technology disciplines. Our authors and editors offer fresh perspectives from universities and industry. Within our subject areas, we have several book series steered by editorial boards made up of leading subject experts.

We peer review each book at the proposal stage to ensure the quality and relevance of our publications.

Get involved

If you are interested in becoming an author, editor, series advisor, or peer reviewer please visit https://www.theiet.org/publishing/publishing-with-iet-books/ or contact author_support@theiet.org.

Discovering our electronic content

All of our books are available online via the IET's Digital Library. Our Digital Library is the home of technical documents, eBooks, conference publications, real-life case studies and journal articles. To find out more, please visit https://digital-library.theiet.org.

In collaboration with the United Nations and the International Publishers Association, the IET is a Signatory member of the SDG Publishers Compact. The Compact aims to accelerate progress to achieve the Sustainable Development Goals (SDGs) by 2030. Signatories aspire to develop sustainable practices and act as champions of the SDGs during the Decade of Action (2020–30), publishing books and journals that will help inform, develop, and inspire action in that direction.

In line with our sustainable goals, our UK printing partner has FSC accreditation, which is reducing our environmental impact to the planet. We use a print-on-demand model to further reduce our carbon footprint.

British Library Cataloguing in Publication Data

A catalogue record for this product is available from the British Library

ISBN 978-1-83953-823-0 (hardback)
ISBN 978-1-83953-824-7 (PDF)

Typeset in India by MPS Limited

Cover image credit: imaginima/E+ via Getty Images

This book is dedicated to
Felix, Daisy, Rafa & Stella

With ceaseless motion comes and goes the tide;
Flowing, it fills the channel vast and wide;
Then back to sea, with strong majestic sweep
It rolls, in ebb yet terrible and deep;
Here samphire-banks and salt-wort bound the flood,
There stakes and sea-weeds withering on the mud;
And higher up, a ridge of all things base,
Which some strong tide has roll'd upon the place.

From *The Borough* by George Crabbe (1810)
A chapter formed by Benjamin Britten into the opera
Peter Grimes (1945),
included in his *Sea Interludes*.

Contents

Preface

This is the 2nd edition of this book, intended to update the important issues of improving wave and tidal generation device reliability and availability.

The development of practical techniques to extract energy from the sea has been the subject of discussion for 200 years.

Real attention has only been paid to wave and tidal device technology in the last six decades, driven by concerns about climate change, the influence of fossil fuel pollution and the risks of energy shortages.

Extracting energy from the tides had been historically applied to the grinding of corn, to sawmills and other energy-intensive agricultural needs, but its application to the mass supply of electricity has only been considered since the Second World War; Hardisty (2009) gives an excellent history.

Practical efforts to extract energy from the waves and tides have received a great deal of attention towards the end of the twentieth century, as concerns about climate change have increased. Over the last 20 years, both these techniques have been demonstrated to be technically viable.

In the last ten years, there have been several practical schemes installed and positive results have been achieved from some of these schemes.

The challenge for the development of marine energy converters (MECs) has been to achieve an acceptable investment cost and rate of return to make them economically viable, compared to other sources of electrical power. The renewable energy resource for MECs is intermittent, so like wind turbines they do not operate at rated power for most of the time. Furthermore, renewable energy converters are exposed to that resource under all these intermittent operational conditions. A key issue of concern to the industry, arising from this intermittency and their exposed situation, has been the reliability and consequent availability of such assets over their lifetime, which needs to be more than ten years and should possibly be greater than 20–30 years. The challenge for the development of MECs will be to achieve an acceptable investment cost and rate of return to make them viable, compared to other sources of electrical power.

This book addresses this problem, by introducing engineers, students and researchers to the issues surrounding reliable design, operation and maintenance of wave and tidal energy devices. By increasing awareness and understanding of the issues facing the industry, it is hoped that reliability can be progressively improved, through better design, testing, deployment and maintenance of these new technologies in their harsh environment. The book identifies wave and tidal power opportunities around the world, studies the development of the devices coming through prototype to productive operation and places them in context. In this respect, for tidal power, I will consider tidal stream energy extraction only and exclude tidal barrage operation, as that is a specialist topic on its own. A sound and consistent reliability theory and taxonomy are developed for wave and tidal devices so that reliabilities can be studied in a comparative way. From this structure, an approach is developed to wave and tidal reliability that can be used to predict the likely availability of these devices coming into service, and their likely life-time performance. This book is not intended to develop the theory of waves or tides and ocean currents or the theory of extraction methods, as these are covered by other texts, for example Cruz Ed (2010) and Hardisty (2009).

Wave and tidal renewable energy generation is still in its infancy. There are many wave and tidal devices under development but as yet very few are actually in revenue earning production or producing useful reliability results. However, it is clear in the United Kingdom, with substantial marine energy resources, that engineering problems are gradually being solved and there is an appetite to invest in renewable generation technologies for harsher environments.

To some extent, wave and tidal is following in the wake of the wind industry, learning from growing off-shore wind farm experience: and it is the author's knowledge from this industry, having studied the reliability, availability and maintainability of off-shore wind turbines, that influences this book. Knowledge in this area is covered in his book *Off-shore Wind Turbines, Reliability, Availability and Maintenance* which acts as a companion. There are some similarities between off-shore wind power and tidal and wave energy extraction: the devices are highly repeatable robotic energy producers, operating full time in the marine environment.

However, wave and tidal energy are clearly different from earlier renewables because, whereas wind turbines are mounted above the sea surface, wave and tidal devices must be submerged below or pierce the sea surface. They are, therefore, more subject to the motion, forces and corrosive effects of the sea. These impose greater technological constraints upon prototype testing, installation, operation and maintainability, which affects their economic viability over a prospective 20–30 year life.

While there are lessons to be learnt from the wind industry, wind turbines (WT) operate in an airflow of 3–25 m/s, whereas wave and tidal MECs operate in a water density 800 times that of air, albeit at a sea-water flow ranging from 0 to 6 m/s. So, for MECs while the renewable energy resource is still intermittent, the resource has a much higher density. However, MECs are exposed to that resource 24 hours per day, seven days per week and must be designed to withstand all the loads the resource imposes under all these intermittent operational conditions. A key issue, of concern to the industry, arising from this intermittency and their exposed situation, has been the reliability and consequent availability of such assets over their lifetime, which needs to be more than ten years and should possibly be 20–30 years. The data presented in this book is based on analytical reliability methods, but the paucity of data sources means that the process is heuristic rather than analytic.

The author first became interested in wave and tidal machine reliability after talking in 2008 to Professor Stephen Salter whose experience, in 1982, of the application of reliability theory to the viability of his Duck wave extraction technology had been frustrating and this author thought there must be a better way to evaluate the prospective reliability of new technology. A methodology for doing this is presented in this book.

Some colleagues have suggested that predicted device failure rates and availabilities, using the methodology in this book, are too pessimistic, perhaps concerned that we may be putting off potential investors.

The author's belief is that these predicted failure rates are based upon heuristic application of proven analytical reliability methods using data from similar environments and that, at this stage in wave and tidal energy development, it is essential for a developing industry to be realistic about the environment's severity, and the demands that will be placed on operational wave and tidal devices throughout their lives. If this can be done now, at the start of this technology, there is potential for engineers and technologists to solve all these problems, designing and operating successful long-lived wave and tidal devices.

In 2010, Professor Stephen Salter of Edinburgh University, inventor of Salter's Duck, said on the development of wave power:

> Every new technology makes many painful mistakes. Many boilers burst, ships sank and planes crashed before we got them reliable. The mistakes only become less painful if people learn from them. They will only learn if full details of every mistake are circulated throughout the industry. That is certainly not happening in this industry now. The requirements for raising private investment require the concealment of expensive disasters, in the hope that commercial rivals will repeat the mistakes.

For this author, Professor Stephen Salter has been a wise, innovative inventor and developer of renewable energy technologies, but he misunderstands the needs of modern energy companies, like Siemens and Oersted, to protect their intellectual property, for them an essential factor in new renewable technology's commercial development.

In 2012, David Shukman, BBC's Science Editor, on the application of MECs said:

> This is an unbelievably harsh environment in which to build anything, let alone manage a vast fleet of machines beneath the waves, there is no rush however: deployments on this kind of scale are at least a decade away, probably more.

This author still agrees with both these sentiments but believes that the time for wave and tidal energy is now upon us.

Peter Tavner, Cambridge, 2024

Acknowledgements

I am indebted to several colleagues who have contributed to this book, most particularly my PhD students Bin Di Chen, Chris Crabtree, Tatiana Delorm, Fabio Spinato, Michael Wilkinson, Mahmout Zaggout and Donatella Zappala.

Also, thanks to undergraduates, research students and postdoctoral research workers who collaborated with me, including Hooman Arabian-Hoseynabadi, Lucy Collingwood, Nathaniel Cowton, Sinisa Djurovic, Yanhui Feng, Rosa Gindele, Andrew Higgins, Mark Knowles, Athanasios Korogiannos, Ting Lei, Luke Longley, Jonathan Moorse, Yingning Qiu, Paul Richardson, Sajjad Tohidi, Wenjuan Wang, Xiaoyan Wang, Matthew Whittle, Jianping Xiang and Wenxian Yang.

I am particularly indebted to my academic colleagues in the Durham Energy Institute, Jim Bumby, Rob Dominy, Simon Hogg, Hui Long, Li Ran and Phil Taylor.

Also, to members of the Supergen Wind Consortium of which I was a member, Mike Barnes, Geoff Dutton, Bill Leithead, Sandy Smith and Simon Watson.

And, to members of the Supergen Marine Consortium of which I was a member, Ian Bryden, Markus Mueller, Stephen Salter, Philipp Thies and Robin Wallace.

And to members of the European Academy of Wind Energy, Berthold Hahn, Stefan Faulstich, Joachim Peinke, Gerard van Bussel and others, who opened my eyes to the potential development of renewable energy throughout Europe.

Finally to Dr R A McMahon, now of Warwick University, for material in Chapter 2.

I would also like to express appreciation for research funding from UK Research & Innovation (UKRI), in Supergen Wind Phases 1 and 2, Supergen Marine Stage 2 and to the European Union EU Framework Programme 7 ReliaWind Consortium, which made the first Edition of this book possible.

I am particularly indebted to my daughter, Sarah Pyne, for proofreading my work and persuading me to clarify the more obscure aspects of this complex and embryonic subject and to Chris Orton of Durham University, Cartographic Unit for carefully preparing all the drawings.

Finally, I would like to thank several industrial organisation colleagues, who assisted by providing data or photographs, including ZF Transmissions, ORE Catapult UK, EMEC UK and Wolong Laurence Scott.

Connectivity Wave Frontispiece, courtesy Michael Carmichael

Nomenclature

Symbol	Explanation
A	Availability, $A = MTBF/(MTBF + MTTR)$, %
A_s	Swept area of an MEC device, m^2
$A(t)$	Availability function of a population of sub-assemblies, as a function of time, $\%(t)$
Acc	Acceleration factor for accelerated life testing
AEP	Annual energy production, MWh
β	Shape parameter in the Weibull probability density function
C	Capacity or load factor, %
CoE	Cost of energy, £/MWh
C_m	Multiplier representing uncertainties associated with a modification method
$COVc_m$	Coefficient of variation for a given C_m
$f(t)$	Probability of failure of a part as a function of time, the probability that the part will fail before time t
FCR	Fixed charge rate for interest, %
FRE	Failure rate estimation process
FRE_{con}	Failure rate estimation, conservative not environmentally adjusted
FRE_{env}	Failure rate estimation, environmentally adjusted
j	Number of components with the same λ
$h(t)$	Hazard or instantaneous failure rate, defined as a limit to failure intensity rate when time differences approach zero. In this book: $h(t) \equiv \lambda(t)$
H	Pressure head of the fluid, m
H_s	Wave height for sea-state, m
η	Efficiency, %
ICC	Initial capital cost, £
I	Drive-train inertia, $kg\text{-}m^2$
i	Part or sub-assembly of a MEC
K_n	Turbine selection coefficient
k	Constant in power balance equations

$\lambda(t)$	Failure intensity or hazard function. Can be represented by a PLP or Weibull function, failures/year
λ	Failure rate. Failure intensity when hazard function is constant with time, failures/year
λ_{ob}	Obtained failure rate
$\lambda_{0.05}$	Prior distribution failure rate, 5% confidence limit
$\lambda_{0.95}$	Prior distribution failure rate, 95% confidence limit
λ_B	Single branch total failure rate, B, the uninterrupted electrical supply assembly
λ_{tot}	Total estimated failure rate, series or parallel network
λ_{tot_Np}	Total estimated failure rate, single branch, N, parallel network
λ_{tot_Ns}	Total estimated failure rate, single branch, N, series network
λ_i	ith sub-assembly estimated failure rate
λ_{i_B}	ith sub-assembly estimated failure rate, in branch B
λ_{i_FREcon}	ith sub-assembly estimated failure rate, conservative not environmentally adjusted
λ_{i_FREenv}	ith sub-assembly estimated failure rate, environmentally adjusted
λ_{Si_min}	ith sub-assembly estimated failure rate, minimum surrogate
λ_{Si_max}	ith sub-assembly estimated failure rate, maximum surrogate
λ_{Gi_min}	ith sub-assembly estimated failure rate, minimum generic
λ_{Gi_max}	ith sub-assembly estimated failure rate, maximum generic
$\mu(t)$	Rate of change of expected number of failures, repair rate, repairs/year
N	Speed of a machine rotor, rev/min
N_{tot}	Total number of device sub-assemblies
$N_{\lambda\beta}$	Number of sub-assemblies in series branch of an assembly, B
N_f	Number of failure events
N_p	Number of sub-assemblies in series within an identical parallel branch
N_s	Number of sub-assemblies in series with none in parallel
n	Estimated device total operating time off-shore, years, for determining λ
P	Device power output, MW
P_{det}	Probability of detection of a fault
p	Integer number of pole pairs
π_{Ei}	Environmental adjustment factor for failure rates
π_{Tbi}	Ocean turbulence adjustment factor for failure rates
π_{Cbi}	Corrosion adjustment factor for failure rates
π_{Hi}	Human adjustment factor for failure rates

Q	Heat flow, Watt/m^2
R	Resistance, Ohms
$R(\lambda, t)$	Reliability survivor function. Hazard function $\lambda(t)$ not constant with time, failures/machine/yr
$R(t)$	Reliability survivor function. The probability that a part will perform its intended function for a specified interval under stated conditions. Hazard function $\lambda\beta$ constant with time, failures/machine/yr
$R_i(t)$	Reliability survivor function of the *ith* sub-assembly
R_a	Model uncertainty. Analytical data
R_r	Model uncertainty. Real-life data
ρ	Fluid density, kg/m^3
r	Discount rate, %
S	Specific energy yield, MWh/m^2/year
σ	Wind speed standard deviation
T	Torque, Nm
T	Period of a wave, sec, or of a tidal cycle, hour
T	Temperature, °C
ΔT	Temperature rise, K or °C
u	Wind speed, m/s, mile/hr, knot
u_r	Rated tidal flow velocity, m/s
u_{sp}	Spring tidal flow velocity, m/s
u_{np}	Neap tidal flow velocity, m/s
u_{peak}	Sea surface velocity, peak, m/s
θ	MTBF of a sub-assembly, hour
V	RMS voltage, V
V_{ref}	Mean wind speed at WT hub height, m/s
W	Work done in a WT Drive-train, MWh
ω	Drive-train angular frequency, rad/s
X_{mod}	Model uncertainty modifier

Abbreviations

Abbreviation	Explanation
AC	Alternating current
ADCP	Acoustic Doppler current profiler
AEP	Annualised energy production
AIP	Artemis Innovative Power
ALT	Accelerated life testing
AM	Asset management
BDFIG	Brushless doubly fed induction generator
BMS	Blade monitoring system
BOP	Balance of plant
CAPEX	Capital expenditure costs
CBM	Condition based maintenance
CMS	Condition monitoring system
CoE	Cost of energy
DC	Direct current
DCS	Distributed control system
DDT	Digital drive technology
DDPMG	Direct drive permanent magnet synchronous generator
DDWRSGE	Direct drive wound rotor synchronous generator and exciter
DE	Drive end of generator or gearbox
DFIG	Doubly fed induction generator
DSD	Device site developers
EAWE	European Academy of Wind Energy
EFC	Emergency feather control
EMEC	European Marine Energy Centre, UK
EPRI	Electric Power Research Institute, USA
EWEA	European Wind Energy Association
FBD	Functional block diagram
FBG	Fibre Bragg grating
FCCC	UN Framework Convention on Climate Change
FCR	Fixed charge rate (interest rate on borrowed money)

FFT	Fast Fourier transform
FM	Field maintenance
FMEA	Failure modes and effects analysis
FMECA	Failure modes, effects and criticality analysis
FSV	Field support vessel
FTA	Fault-tree analysis
GB	Ground benign: protected environment
GF	Ground fixed: severe environment
GHG	Greenhouse gas
GM	Ground mobile environment
HM	Health monitoring
HPP	Homogeneous Poisson process
H&S	Health and safety
HSS	Gearbox High speed shaft
HV	High voltage
ICS	Integrated control system
IDFTlocal	Iterative localised discrete Fourier transform
IEA	International Energy Agency
IEC	International Electrotechnical Commission
IEEE	Institution of Electrical and Electronic Engineers, USA
IET	Institution of Engineering and Technology (formerly IEE)
IG	Induction generator
IM	Information management
IMS	Gearbox intermediate shaft
IP	Intellectual property
ISET	Institut für Solare Energieversorgungstechnik, Kassel, Germany
LCC	Life cycle costing
LSS	Gearbox low speed shaft
LV	Low voltage
LWK	Landwirtschaftskammer Schleswig-Holstein (wind turbine database for Germany)
MCA	Marine and Coastguard Agency
MEC	Marine energy converters
MIL-HDBK	US Reliability Military Handbook
MM	Maintenance management
MTBF	Mean time between failures
MTTR	Mean time to repair
MV	Medium voltage

NAREC	National Renewable Energy Centre, UK
NDE	Non-drive end of generator or gearbox
NHPP	Normal homogeneous Poisson process
NOAA	National Oceanic and Atmospheric Administration, USA
NPRD	Non-electronic parts reliability data
NS	Naval sheltered: normal environment
NU	Naval unsheltered: severe environment
OEM	Original equipment manufacturer
OFGEM	Office of gas and electricity markets
OFTO	Off-shore transmission owner
OM	Operations management
O&M	Operations and maintenance
OPEX	Operational expense costs
ORE	Off-shore renewable energy catapult, owner of NAREc
OREDA	Off-shore and on-shore reliability data
OWT	Off-shore wind turbine
PCRPT	Parts count reliability prediction technique
PDF	Probability distribution function
PLC	Programmable logic controller
PLP	Power law process
PMG	Permanent magnet generator
PMSG	Permanent magnet synchronous generator
PSD	Portfolio of surrogate data
PSD	Power spectral density
PSF	Premature serial failure of a sub-assembly
PTO	Power take-off unit in a WEC or TSD
RB	Reliability block
RBD	Reliability block diagram
RMP	Reliability modelling and prediction
RNA	Rotor nacelle assembly
RPN	Risk priority number
SBPFgear	Sideband power factor
SCADA	Signal conditioning and data acquisition
SCIG	Squirrel cage induction generator
SVC	Static VAr compensator
TBF	Time between failures
TDRM	Time-dependent reliability model
TSD	Tidal stream device

TTF	Time to failure
TTT	Total time on test
VGB	German Power Standards Company
VSC	Voltage source converter
WEC	Wave energy converter
WF	Wind farm
WMEP	Wissenschaftlichen Mess- und Evaluierungsprogramm data- base at Kassel for Germany
WRIG	Wound rotor induction generator
WRIGE	Wound rotor induction generator and exciter
WRSGE	Wound rotor synchronous generator and exciter
WSD	Windstats database for Germany
WSDK	Windstats database for Denmark
WT	Wind turbine

About the author

Peter Tavner, Eur Ing, CEng, FIET, is Emeritus Professor in the Department of Engineering, Durham University, UK. He received an MA in Mechanical Sciences from Cambridge in 1970, a PhD from Southampton in 1978 and a DSc from Durham in 2012.

He served as a Royal Navy officer from 1964 to 1971 in General Service on HM Ships ASHANTI, BRISTOL and LONDON, subsequently working in a Medical Research Council laboratory and then teaching at Benin University, Nigeria from 1972 to 1974.

He has held a number of senior research and technical positions in the electrical supply and manufacturing industries from 1979 to 2002, including Group Head in a former CEGB research laboratory, Technical Director of Laurence, Scott & Electromotors Ltd (LSE) then of Brush Electrical Machines Ltd (BEM), two of the UK's large electrical machine manufacturers, becoming Group Technical Director of FKI Energy Technology, owner of LSE & BEM, an international business manufacturing wind turbines, electrical machines and drives in the United Kingdom, Holland, Italy, Germany and the Czech Republic.

He won with Richard Jackson an IEE Institution Premium in 1988 for a paper on discharge current coupling between electrical machine conductors.

He was Professor of New & Renewable Energy at Durham University from 2003 to 2011, establishing MEng and MSc courses in New & Renewable Energy in the Department placing it as a leading UK centre for off-shore renewable technologies, being Head of Department from 2006 to 2010.

He was President of the European Academy of Wind Energy from 2010 to 2012 and won the European Academy of Wind Energy Scientific Award in 2020 for a pioneering role in research into wind farm reliability, availability and maintenance, being made a Life Member of the Academy.

He is the author of the following four books published by the IET:

- *Condition Monitoring of Rotating Electrical Machines* with Profs Li Ran and Chris Crabtree, now in its 3rd Edition, 2020.
- *Offshore Wind Turbines: Power, reliability, availability and maintenance* in its 2nd Edition, 2021.
- *Clean Energy: Past to future* in its 1st Edition, 2023.
- *Wave and Tidal Generation Devices: Reliability and availability* in its 2nd Edition, 2024.

Chapter 1
Overview of wave and tidal development

1.1 Introduction

The challenge for the development of marine energy converters (MEC) is to achieve an investment cost and rate of return that will make marine energy devices viable, compared to other sources of electrical power.

While marine energy resources are of reasonable density, they are intermittent and MECs do not operate at fully rated power for most of their time, despite being exposed to the resource 24 hours a day, 7 days a week.

Before we can address, in detail, the challenges that this presents, we must first have an overview of wave and tidal developments to date.

1.2 World wave resource

Wave energy in the ocean is transferred from solar energy by the wind (Cruz Ed 2010).

The input solar flux averages 350 Watts/m^2, whereas the power density in the waves from the wind is typically 0.01–0.1 Watts/m^2.

So, the driving force is weak, but the huge fetch of ocean waves can lead to much greater energy front densities of 100 kW/m in some parts of the ocean (Barstow *et al.* 2003).

Information from buoys and satellite imaging has been used to estimate the energy available in waves (Cruz Ed 2010), but this depends on synoptic weather conditions, the coastline plan geometry and the inclination of the shore.

Wave buoys have been used since the 1960s and a number of wave buoy types, such as the Waverider, using the Hippy 40 sensor, and the Wavescan, have proved reliable over periods of more than 20 years, an important lesson for MEC designers.

Wave buoys are generally deployed as part of national energy programmes, and the most extensive deployments are by the United States, NOAA-NDBC, the Indian National Databuoy Programme, some national networks in Spain, Greece and Italy. There are also wave-buoy deployments in the Irish and North Seas from Ireland and the United Kingdom up to Norway, in support of the present and future off-shore oil and gas industry.

Since 1980, radio altimetry from a variety of satellites has been used to determine the heights of waves, with wind information interpretable from Doppler back-scatter. This is available at all seasons of the year in all parts of the globe, but with greater accuracy as the satellite tracks move closer to the tropics, generating vast quantities of wave height and wind data. The sparsely distributed but accurate wave buoy networks have provided a reliable way of verifying and calibrating data from satellite altimetry. In this way, several important comparative studies have strengthened the reliability of the much more numerous and detailed satellite altimetry data.

From this work, several wave climate models, relying on computer hindcasting of wind-wave models, have been developed. These include the WAM model, which has led to the development of the European Wave Energy Atlas (WERATLAS), WAMDI Group (1988), which in turn lead to a long-term programme to develop an integrated MATLAB package, WorldWaves. This calculates energy time series from directional wave spectra and associated statistics for coastal waters anywhere in the world, enabling developers to make predictions from modelled information, based on satellite altimetry data, verified by wave buoy measurements.

A very broad summary of world wave energy results, close to shore, is shown in Figure 1.1, summarised as the average power per metre length of wave front during a one-year period. It is currently believed that economically viable power generation can be secured from wave fronts with a linear energy density greater than 30 kW/m.

1.3   World tidal resource

Tidal activity is recorded in many sea-bound countries using tidal gauges, the data being transmitted back to a central point via landline, radio or satellite.

The difficulty when considering global resource is to gather this disparate data from many different countries, which use widely varying measurement systems. The geometry of the coastline, the relative proximity of neighbouring coasts and the prevailing synoptic weather conditions all affect tidal heights and their energy resources.

Again, satellite imaging information can be used to give a global perspective (Cruz Ed 2010), but the data provided can only give an indication of the relative geographical intensity of the resource. Local measurements are essential to confirm the validity of any local scheme, taking account of local coastline and currents. Figure 1.2 identifies the locations where the tides and coastal geometry favour the extraction of energy from the tidal stream. The energy density of tidal resource is more difficult to plot than wave and wind since it is much more sensitive to the exact geographical location and the type of device installed in the tidal stream.

1.4   Influence of the weather

The ambient weather conditions have a large effect on wave device power outputs, but a less significant effect on tidal devices, because of the smoothing effect in the

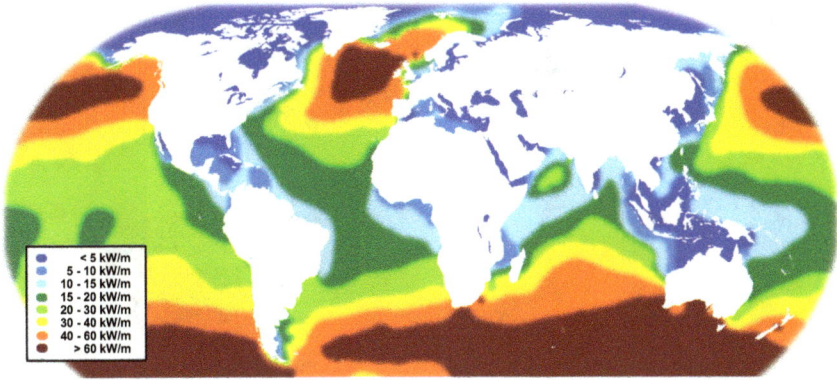

(a)

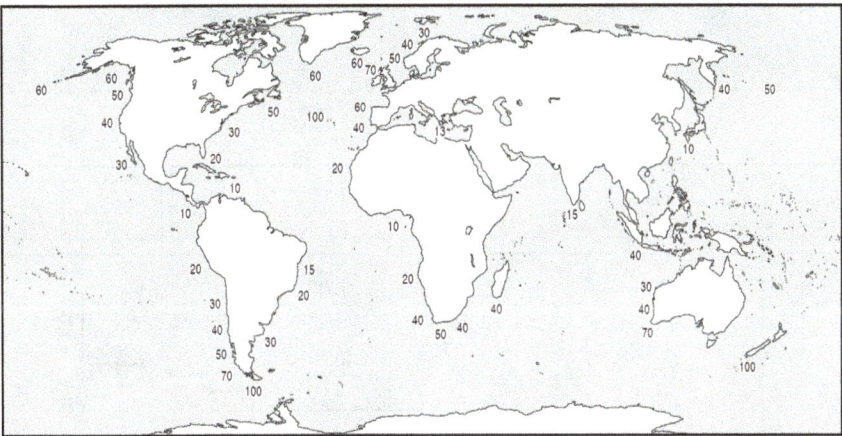

(b)

Figure 1.1   Wave energy resource, close to shore; world geographical distribution: (a) visual in kW/m: Source: Exowave and (b) numerical in kW/m: Source: Cruz Ed

water column. The latter can be affected when the wind acts in the same direction as the tidal flow.

These effects can, however, play an important part in the maintainability of both wave and tidal devices because they affect their performance, deterioration and most importantly their access. The weather, therefore, can have a large effect on wave and tidal device availability.

The Beaufort scale (Wikipedia 2012) shown in Table 1.1 was devised in 1805 to provide a simple measure of wind and sea conditions for shipping but is now used in the study of ocean renewable energy systems.

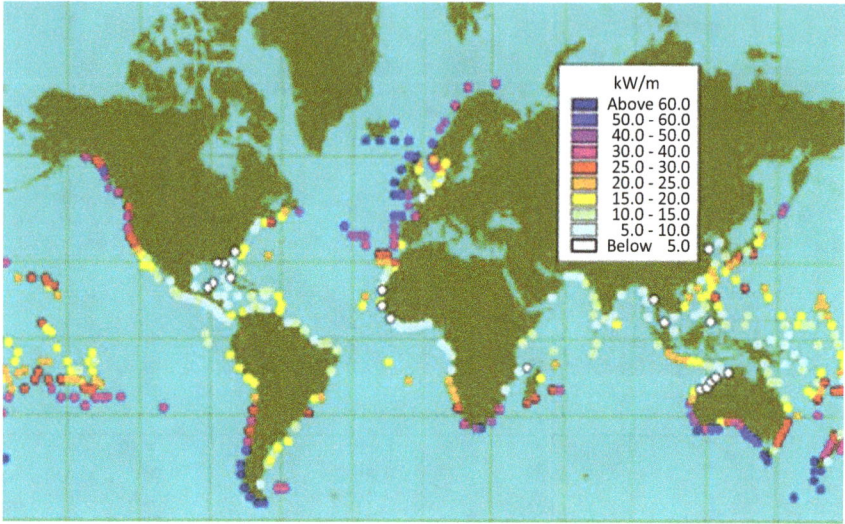

(a)

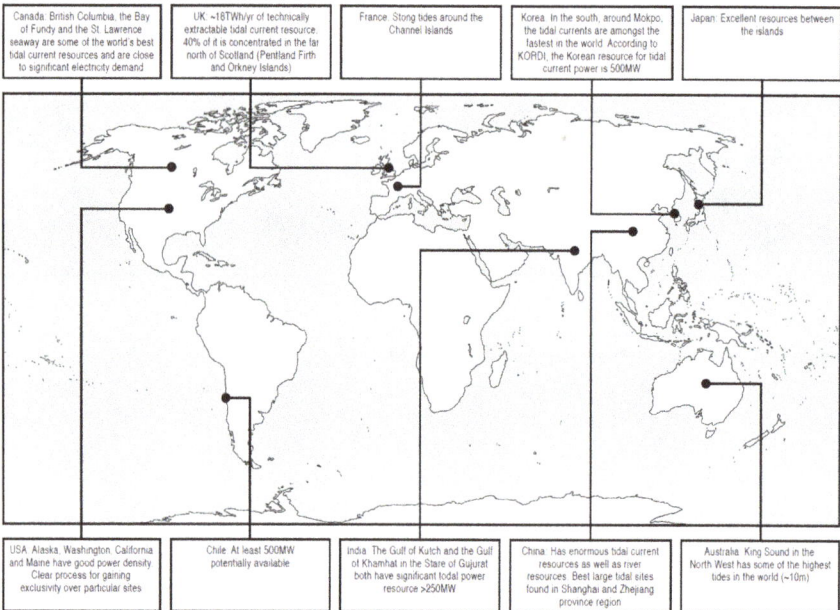

| Canada: British Columbia, the Bay of Fundy and the St. Lawrence seaway are some of the world's best tidal current resources and are close to significant electricity demand | UK: ~18TWh/yr of technically extractable tidal current resource. 40% of it is concentrated in the far north of Scotland (Pentland Firth and Orkney Islands) | France: Stong tides around the Channel Islands | Korea: In the south, around Mokpo, the tidal currents are amongst the fastest in the world. According to KORDI, the Korean resource for tidal current power is 500MW | Japan: Excellent resources between the islands |

| USA: Alaska, Washington, California and Maine have good power density. Clear process for gaining exclusivity over particular sites | Chile: At least 500MW potentially available | India: The Gulf of Kutch and the Gulf of Khamhat in the Stare of Gujurat both have significant todal power resource >250MW | China: Has enormous tidal current resources as well as river resources. Best large tidal sites found in Shanghai and Zhejiang province region | Australia: King Sound in the North West has some of the highest tides in the world (~10m) |

(b)

Figure 1.2   *Tidal resource close to shore; world geographical distribution:*
(a) global renewable tidal energy resources and (b) promising tidal
locations, identified by Krogstad et al. (1999)

Table 1.1   Beaufort scale for weather at sea

Beaufort number	Description	Wind speed	Wave height	Sea conditions
0	Calm	< 0.3 m/s	0 m	Flat
1	Light air	0.3–1.5 m/s	0–0.2 m	Ripples without crests
2	Light breeze	1.6–3.4 m/s	0.2–0.5 m	Small wavelets. Crests of glassy appearance, not breaking
3	Gentle breeze	3.4–5.4 m/s	0.5–1 m	Large wavelets. Crests begin to break; scattered whitecaps
4	Moderate breeze	5.5–7.9 m/s	1–2 m	Small waves with breaking crests. Fairly frequent whitecaps
5	Fresh breeze	8.0–10.7 m/s	2–3 m	Moderate waves of some length. Many whitecaps. Small amounts of spray
6	Strong breeze	10.8–13.8 m/s	3–4 m	Long waves begin to form. White foam crests are very frequent. Some airborne spray is present
7	High wind, moderate gale, near gale	13.9–17.1 m/s	4–5.5 m	Sea heaps up. Some foam from breaking waves is blown into streaks along wind direction. Moderate amounts of airborne spray
8	Gale, fresh gale	17.2–20.7 m/s	5.5–7.5 m	Moderately high waves with breaking crests forming spindrift. Well-marked streaks of foam are blown along wind direction. Considerable airborne spray
9	Strong gale	20.8–24.4 m/s	7–10 m	High waves whose crests sometimes roll overDense foam is blown along wind direction. Large amounts of airborne spray may begin to reduce visibility
10	Storm, whole gale	24.5–28.4 m/s	9–12.5 m	Very high waves with overhanging crests. Large patches of foam from wave crests give the sea a white appearanceConsiderable tumbling of waves with heavy impact. Large amounts of airborne spray reduce visibility
11	Violent storm	8.5–32.6 m/s	11.5–16 m	Exceptionally high waves. Very large patches of foam, driven before the wind, cover much of the sea surfaceVery large amounts of airborne spray severely reduce visibility
12	Hurricane force	≥ 32.8 m/s	≥ 16 m	Huge wavesSea is completely white with foam and sprayAir is filled with driving spray, greatly reducing visibility

Rows 0–10 of Table 1.1 highlight in dark, medium and light grey are the ranges of weather conditions in which off-shore wind turbines can operate continuously.

Whereas rows 0–9 of Table 1.1 highlight in dark and medium grey are the ranges of weather conditions in which wave and tidal devices could be reasonably expected to operate continuously.

The dark grey area, i.e. rows 0–5, is a more restricted range, with waves up to 3 m, in which specialised vessels can access wind, wave or tidal devices for maintenance purposes; at higher wave heights, the devices then become inaccessible to seaborne support.

1.5   Tidal power

1.5.1   Device development

Tidal power is the most straightforward of marine renewable energy sources; the resource is predictable and the locations of highest tidal power energy are clearly identifiable, although relatively few in number.

We have identified the marine locations providing the strongest tidal resources, from which we effectively have three technical choices:

- To impound the tidal area, enclosing it in a barrage, which the sea-water can flow through installed turbines.

  This approach requires the application of large and costly civil engineering works but has the advantage of longevity, provided the installed turbines can survive the barrage life, or be readily replaced in situ.
- To install turbines on the sea bottom, located in areas of strong tidal flow.

  This has the advantage of simplicity but requires the deployment of a large number of complex devices in the marine environment, requiring underwater access for regular maintenance, in difficult conditions, with uncertainty over the device life-spans.
- To install moored turbines, floating on the sea surface in areas of strong tidal flow.

  This approach has the advantage that devices can be unmoored and towed to harbour for maintenance or repair, which will almost certainly lead to longer device life expectancy.

Table 1.2, like that presented by Delorm (2014), is developed from a list given by EMEC and displayed in Wikipedia, showing the large range of tidal stream devices (TSD) being considered and the countries where that development is taking place.

The longer length of the list in this new edition reflects the predictability of tidal power, which is attracting investors; however, lower capital costs and in-service operational reliability are still issues which new developers must demonstrate.

Table 1.2 Tidal stream devices operational and planned

Station	Capacity, MW	Turbines	Country	Commissioned or start
Operational				
Rance Tidal Power Station	240	24 × 10 MW reversible Kaplan turbines	France	1966 renovated 2011
Kislaya Guba Tidal Power Station	1.7	0.4 MW French-built unit. new 0.2 MW unit, 2nd 1.5 MW unit	Russia	1968 renovated 2004 and 2007
Haishan Tidal Power Plant	0.25		China	1975
Jiangxia Tidal Power Station	4.1	1 × 600 kW, 5 × 700 kW	China	1980
EMEC Fall of Warness tidal test site	up to 10	1 × 2 MW Orbital O2 1 × Magallanes Renovables ATIR	UK, Scotland	2007
Uldolmok Tidal Power Station	1.5	200 kW	South Korea	2009
Sihwa Lake Tidal Power Station	254	10 × 25.4 MW bulb turbines	South Korea	2011
Bluemull Sound Tidal Stream Array	0.3	3 × 100 kW Nova Innovation M100D	UK, Scotland	2016
MeyGen	6	4 × 1.5 MW	UK, Scotland	2017
Minesto Vestmannasund	1.4	1 × 1.2 MW, 2 × 100 kW tidal kites	Faroe Islands	2022, 2024

(Continues)

Table 1.2 (*Continued*)

Station	Capacity, MW	Turbines	Country	Commissioned or start
Under construction, following list of projects construction as of the date displayed.				
Morlais West Anglesey Demonstration Zone	240 MW	No details	UK, Wales	Consented 2021 1st tidal device 2026
Proposed, following list of projects at a proposal stage, schemes may not go ahead, but have not formally been cancelled.				
EURO-TIDES project	9.6	4 × Orbital O2	UK, Fall of Warness, Orkney Scotland	-
Garorim Bay Tidal Power Station	520		South Korea, Garorim Bay	-
Gulf of Kutch Project	50		India, Gulf of Kutch	-
Incheon Tidal Power Station	818 or 1,320		South Korea	-
Mezenskaya Tidal Power Plant	24,000		Russia, Mezen Bay	-
Penzhin Tidal Power Plant Project	89,100		Russia, Penzhin Bay	-
Seastar project	4	16 × 250 kW Nova Innovation	UK, Fall of Warness, Orkney, Scotland	-
Severn Barrage	8,640		UK, Severn Estuary	-
Tidal Lagoon Swansea Bay	320		UK, Swansea Bay, Wales	-
Tugurskaya Tidal Power Plant	3,640		Russia, Okhotsk Sea	-

1.5.2   Tidal device categorisation

Tidal devices may have the following characteristics:

Tidal lagoons or barriers:
- Energetic tidal locations around the world can be exploited by impounding an area using a surrounding barrier to create a lagoon or as proposed by dynamic tidal power employing barriers extending from the shore;

Moored and floating devices:
- Single- or double-axis horizontal turbines, driving generators directly or through a gearbox. Supply to the shore is via floating cables to a sea-bed termination and then trenched, sub-sea electrical cables. The floating arrangement allows a device to be detached and removed for maintenance.

Fixed monopile devices:
- Single- or double-axis horizontal turbines, driving generators directly or through a gearbox, with the whole mounted on a fixed monopile in the sea-bed.
- Supply to the shore is via a J-tube in the monopile and trenched, sub-sea electrical cables.

Sea-bed devices:
- Lowered onto and fixed to the sea bottom, these carry horizontal or vertical axis turbines that drive a generator directly or through a gearbox, all mounted on a gravity-based foundation.
- Supply to the shore is via trenched, sub-sea electrical cables.

Tidal turbines can be open flow or ducted to improve their flow geometry.

Tidal turbine generator electrical outputs can be synchronised to the mains alternating current (AC) by inserting a voltage source converter (VSC) between the generator and grid connection. This converter may be mounted within the floating or fixed off-shore structure or be placed on-shore at the end of the trenched, sub-sea electrical cables.

Table 1.2 lists various TSDs under development and the limited number of categories demonstrates less variance in the scope of designs proposed, compared with wave energy converters (WECs).

There are, however, many proposed prototypes, perhaps reflecting greater investor confidence in the predictability of the tidal resource.

It is highly likely that numerous and diverse list presented in Table 1.2 will become simplified as experience grows in the industry.

Therefore, it should be more straightforward to identify configurations likely to be reliable. We are still in a relatively early stage of development where, again, there is a lack of consensus on the most applicable or reliable extraction technology, hence the motivation of this book.

Figure 1.3 with one tidal lagoon and three other tidal current prototypes demonstrates that a convergence of consensus is appearing in major operational technologies.

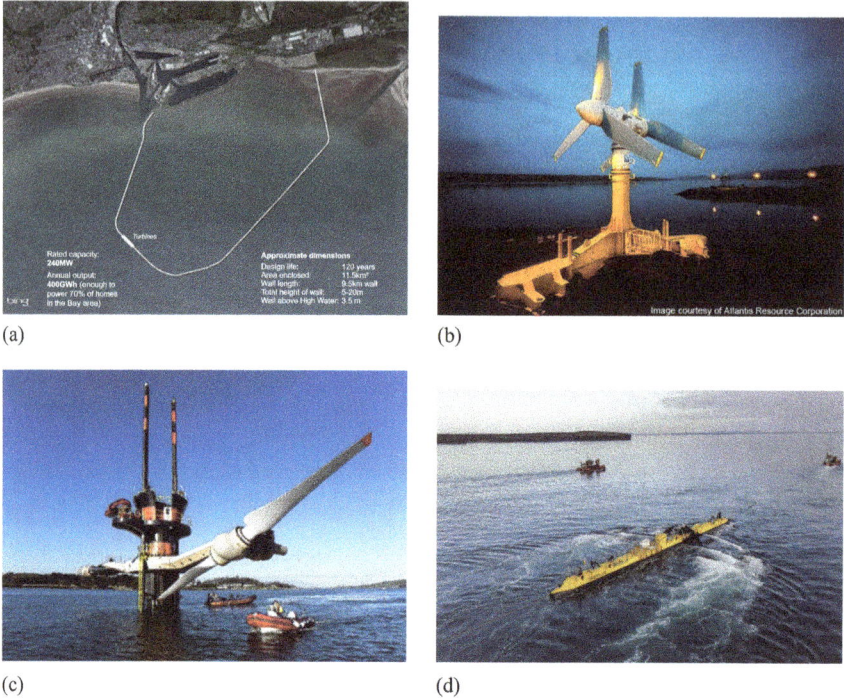

(a)   (b)

(c)   (d)

Figure 1.3   *Four TSD technologies tested or considered in last 15 years, taken*
from Table 1.2. (a) Image of a proposed tidal lagoon in Swansea Bay,
UK, (b) Atlantis AK1000, tidal device fixed to the sea-bed, (c) Seagen,
raisable tidal device fixed to the sea-bed in Strangford Lough, UK,
(d) Orbital O2, floating tidal device moored to the sea-bed in the
Fall of Warness, UK

1.6   Wave power

1.6.1   *Device development*

Wave power devices are more complex than other marine renewables, because of
the complexity of the incoming resource and the individuality of nearshore marine
environments, which are a very tough location for deployment.

Consensus has not yet been reached on the most favourable technologies to
apply, but many options are available, as this book will show.

1.6.2   *Wave device categorisation*

Wave power extraction is more complex than tidal power, partly because the
resource is more point powerful than the tide but primarily because the resources
are distributed and cannot easily be aggregated.

We can identify marine locations that provide the strong tidal resource from which we effectively have three technical choices;

- To impound the tidal area, enclosing it in a barrage which the sea-water can flow through installed turbines.

  This approach requires the application of large and costly civil engineering works but has the advantage of longevity, provided the installed turbines can survive the barrage life, or be readily replaced in situ;

- Install turbines on the sea bottom in areas of strong tidal flow.

  This has the advantage of simplicity but requires the deployment of a large number of complex devices in the marine environment, requiring underwater access for regular maintenance with uncertainty over the devices lifespan;

- Install moored turbines floating on the surface of the sea in areas of strong tidal flow.

  This has the advantage that devices can be unmoored and towed to harbour for maintenance and repair, which will almost certainly lead to longer life expectancy.

Table 1.2, taken from Delorm (2014), is developed from a list given by EMEC and displayed in Wikipedia, showing the large range of TSD being considered and the countries where that development is taking place.

The length of the list reflects the predictability of tidal power, which attracts investors; however, operational reliability in service has proved a problem.

The European Marine Energy Centre (EMEC) in the United Kingdom has identified eight WEC device categories that can be developed for wave energy extraction:

- Surface following attenuators:

  Moored to the seabed and floating in the direction of wave travel, these absorb energy through hydraulically driven generators powered by the advancing wave. Supply to the shore is via floating cables to a sea-bed termination and then via trenched electrical cables.

  A version of this approach, Bulge Technology, consists of a rubber tube filled with water. The water enters through the stern and the passing wave causes pressure variations along the length of the tube, creating a 'bulge'.

  As the bulge travels through the tube it grows, gathering energy which can be used to drive a standard low-head turbine located at the bow, where the water then returns to the sea.

- Point absorbers or buoys:

  Moored, floating structures that absorb energy from all directions through their movements at/near the water surface. It converts the motion of the buoyant top relative to the base into electrical power. Supply to the shore is via floating cables to a sea-bed termination and then via trenched electrical cables. The power take-off systems take different forms, depending on the configuration of displacers/reactors.

- Multi-point absorbers:
  Combinations of a number of point absorbers in an array.
- Oscillating wave surge converters:
  These extract energy from wave surges and the movement of water particles within them. The arm oscillates as a pendulum mounted on a pivoted joint, in response to the movement of water in the waves, and energy is extracted through hydraulically driven generators. Supply to the shore is via floating cables to a sea-bed termination and then via trenched electrical cables.
- Oscillating water columns:
  Partially submerged, hollow structures, these are open to the sea, below the water line, enclosing a column of air on top of a column of water. Waves cause the water column to rise and fall, which in turn compresses and decompresses the air column. This trapped air is allowed to flow to and from the atmosphere via a turbine and generator, which usually has the ability to rotate regardless of the direction of airflow. The rotation of the turbine is used to generate electricity on-shore, and supply is then via trenched electrical cables.
- Overtopping or termination devices:
  These capture water as waves break into a moored, floating storage reservoir. This water is then returned to the sea, passing through a conventional low-head turbine that generates power. Supply to the shore is via floating cables to a sea-bed termination and then via trenched electrical cables. An overtopping device may use collectors to concentrate the wave energy.
- Submerged pressure differential or underwater attenuator devices:
  Typically moored near the shore but attached to the seabed, the motion of the waves causes the sea level to rise and fall above the device. This induces a pressure differential within the device, which either pumps fluid through a system to generate electricity or directly drives a linear generator. Supply to the shore is via trenched electrical cables.
- Rotating mass:
  Floating technology in which an unbalanced mass rocks about a vertical axis and energy is extracted as the wave rotates the mass about that axis driving a rotational generator. Supply to the shore is via floating cables to a sea-bed termination and then via trenched electrical cables.

1.6.3   Device development

Table 1.3, taken from Wikipedia (2023), shows the wide range of WECs being developed around the world, the technologies involved and the countries where that

Table 1.3   WECs under recent development

Project	Developer	Location	Technology	Site	Distribution	Operation	Description
Azura wave power device	US Navy Wave Energy Test Site	Kaneohe Bay, Hawaii, USA	Submerged	Off-shore	Electric	2016	45-ton wave energy converter located at a depth of 30 metres.
Albatern WaveNET	Albatern	Scotland UK	Multi-point absorber array	Off-shore		2010	Albatern are working on their third iteration devices with a 14-week deployment on a Scottish fish-farm site in 2014 and a six-unit array deployment for full characterisation at Kishorn Port in 2015.
AMOG, AEP WEC	Falmouth	Cornwall UK	Surface dynamic vibration absorber	Off-shore	Electric	2019	Operation off SW England, 1/3rd scale device successfully deployed in Europe summer 2019 at FaBTest. Device's maximum rating is 75 kW.
Anaconda Wave Energy Converter	Checkmate SeaEnergy		Surface-following attenuator	Off-shore	Hydro-electric turbine	2008	Early stage development of a 200 m long rubber tube device, tethered underwater. Passing waves instigate a wave inside the tube, which then propagates down its walls, driving a far-end turbine.
Aqua Buoy	Finavera Wind Energy, later SSE Renewables	Ireland–Canada–Scotland	Buoy	Off-shore	Hydro-electric turbine	2003	In July 2010, Finavera Renewables announced that it had entered a definitive agreement to sell all assets and Intellectual Property related to the AquaBuOY wave energy technology.
Atmocean	Atmocean Inc.	USA	Point Absorber array	Nearshore and off-shore	Pump-to-shore	2006	Atmocean 15, 3 m diameter surface buoys anchored at six points.

(Continues)

Table 1.3 (Continued)

Project	Developer	Location	Technology	Site	Distribution	Operation	Description
							Each buoy uses passing waves to pump sea-water on-shore where it enters a desalination process without need of external energy. Advantage of smaller modular system includes using standard shipping containers and small boat operations. Two full-scale trials deployed off coast of Ilo, Perú in 2015.
AWS-III	AWS Ocean Energy	Scotland UK	Surface-following attenuator	Off-shore	Air turbine	2010	AWS-III is floating toroidal vessel with rubber membranes on the outer faces that deform as waves pass, moving air inside chambers to drive air-turbines and generate electricity. Tested in 1/9 scale model in Loch Ness in 2010, and are now working on a full-size version 60 m across to generate 2.5 MW. Envisaged to be installed in off-shore farms moored in 100 m depth of water.
CalWave	CalWave Inc.	USA	Submerged pressure differential	Off-shore		2021	CalWave Power Technologies, Inc commissioned a pilot unit device off the coast of San Diego.
CCell	Zyba Renewables	UK	Oscillating wave surge converter	Nearshore and off-shore	Hydraulic	2015	CCell is a directional WEC consisting of a curved flap operating in the surge direction of wave propagation. Energy is dissipated over a long arc reducing wave height, and its shape cuts through the waves reducing boundary turbulence. In addition, CCell is designed to float just under the water surface, maximising available wave energy.

CETO Wave Power	Carnegie	Australia	Buoy	Off-shore	Pump-to-shore	2008	Being tested off Fremantle, Western Australia and consists of a single piston pump attached to the sea floor with a buoyant float tethered to that piston. Waves cause the float to rise and fall, pressurising water piped to an on-shore facility to drive hydraulic generators or run reverse osmosis water desalination.
Crestwing	Crestwing ApS	Denmark	Surface-following attenuator	Off-shore	Mechanical	2014	Consists of two floats connected by a hinge, uses atmospheric pressure acting on its large surface to adhere to the ocean surface, allowing it to follow wave motion, using its two floats to convert KE and PE energy into electricity via a mechanical power take-off system. In 2014, a 1:5 scale model was tested in the sea near Frederikshavn.
Cycloidal Wave Energy Converter	Atargis Energy Corporation	USA	Fully Submerged Wave Termination Device	Off-shore	Direct Drive Generator	2006	In the tank testing stage of development, the device was a 20 metres diameter fully submerged rotor with two hydrofoils. Numerical studies have shown greater than 99% wave power termination capabilities. These were confirmed by experiments in a small 2D wave flume and large off-shore wave basin.
FlanSea	FlanSea	Belgium	Buoy	Off-shore	Hydroelectric turbine	2010	A point absorber buoy developed for use in southern North Sea conditions. It works by means of the bobbing effect of the buoy on a cable to generate electricity.

(Continues)

Table 1.3 (Continued)

Project	Developer	Location	Technology	Site	Distribution	Operation	Description
HiWave-5	CorPower Ocean	Portugal	Point absorber buoy	Off-shore	Gearbox and generator	2023	300 kW rated power, part of the HiWave-5 array demonstration project.
Islay LIMPET	Islay LIMPET	Scotland UK	oscillating water column	On-shore	Air turbine	1991	500 kW shoreline device using an oscillating water column to drive air in and out of a pressure chamber through a Wells turbine.
Lysekil Project	Uppsala University	Sweden	Buoy	Off-shore	Linear generator	2002	Direct driven linear generator placed on the seabed, connected to a buoy at the surface via a line. The movements of the buoy will drive the translator in the generator.
Neptune Wave Engine	Neptune Equipment Corp.	Vancouver Canada	Multiple Point Absorbers	Near Shore – Small 0.1 to 5 m Waves	Direct Drive Mechanical PTO	2010 Updated 2019	Wave energy captured by multiple float-pistons constrained to move vertically up and down. Reciprocating float-piston motion converts to one-way rotation motion via a patented PTO allowing power to be applied to generator on both up and down strokes. 5 full-size test units deployed and a sixth was deployed in 2019.
Ocean Grazer	University of Groningen	The Netherlands	Buoy	Off-shore	Hydraulic multi-piston pump	2011	Wave energy captured by multiple hydraulic pistons on a floater. Main advantage over other systems is that it adapts itself to any wave shape having a high 70% efficiency.
Oceanlinx	Oceanlinx	Australia	OWC	Nearshore and Off-shore	Air turbine	1997	Wave energy captured by an Oscillating Water Column, with electricity generated by air flowing through a turbine. The third

Name	Company	Country	Type	Location	PTO	Year	Description
							medium scale demonstration unit near Port Kembla, NSW, Australia, a medium-scale system that was grid connected in early 2010. A full scale commercial nearshore unit, *greenWAVE*, with a 1MW capacity will be installed off Port MacDonnell in South Australia by 2013.
Oceanus 2	Seatricity Ltd	UK	Buoy	Nearshore and Off-shore	Pump-to-shore	2007	First device deployed and tested at UK's WaveHub site, 2014–2016. Third-generation device consists of a single piston patented pump mounted on a gimbal and supported by a 12 m diameter buoy/float. Pump tethered to the seabed. Vertical wave motion pumps sea-water at pressure through a pipe to on-shore, driving a hydraulic generator or to run a reverse osmosis water desalination.
OE buoy	Ocean Energy	Ireland	Buoy	Off-shore	Air turbine	2006	The OE Buoy has only one moving part and in 2009 completed a two-year sea trial at quarter-scale. A full-scale version commenced construction in Oregon in 2018 and was scheduled to deploy at the USN Wave Energy Test Centre in 2019.
OWEL	Ocean Wave Energy Ltd	UK	Wave Surge Converter	Off-shore	Air turbine	2013	The surge motions of long-period waves compress air in a tapered duct which is then used to drive an air turbine mounted

(Continues)

Table 1.3 (*Continued*)

Project	Developer	Location	Technology	Site	Distribution	Operation	Description
							on top of a floating vessel. Design of a full-scale demonstration project was completed in Spring 2013, ready for fabrication.
Oyster Wave Energy Converter	Aquamarine Power	UK Scotland Ireland	Oscillating Wave Surge Converter	Near-shore	Pump-to-shore hydro-electric turbine	2005	A hinged mechanical flap attached to the seabed captures nearshore wave energy. Flap drives hydraulic pistons to deliver high-pressure water to an on-shore turbine generating electricity. In 2009, the first full-scale demonstrator began producing power at the European Marine Energy Centre's wave test site at Billia Croo in Orkney.
Pelamis Wave Energy Converter	Pelamis Wave Power	UK Scotland	Surface-following attenuator	Off-shore	Hydraulic	1998	Waves exercise a series of semi-submerged cylinders linked by hinged joints, which move relative to one another, activating hydraulic cylinders pumping pressurised oil through hydraulic motors, driving electric generators. P1 installed at EMEC in Orkney in 2004, becoming the world's first off-shore wave energy device to generate electricity into the National Grid. P2, owned by E.ON, started grid-connected tests off Orkney in 2010. Aguçadoura Wave Farm in Portugal, 2008, first commercial application of Pelamis technology.

Penguin	Wello Oy	Finland	Rotating mass	Off-shore	Direct Conversion	2008	First 0.5 MW device deployed at EMEC in Summer 2012. Unit modified and reinstalled at Billia Croo in 2017, part of the Horizon 2020, funded by Clean Energy From Ocean Waves. CEFOW is a five-year project to deploy 3 off 1 MW Penguin units in off-shore conditions in a grid-connected testing environment. Project is coordinated by utility company Fortum. Wello Penguin deployed in Orkney waters 2014.
R38/50 kW and R115/150 kW	40 South Energy	UK	Underwater attenuator	Off-shore	Electrical conversion	2010	Machines work by extracting energy from relative motion between one Upper Member and one Lower Member following an innovative method which in 2011 earned company a UKTI R&D Award. First reduced-scale prototype tested off-shore during 2007. Second-generation full-scale prototype tested off-shore in 2010. Third-generation full-scale prototype tested off-shore in 2011. First units sold to clients in various countries in 2012. Recognised as one of the technological innovators in the sector.
Sea Power	Seapower Ltd.	Ireland	Surface-following attenuator	Off-shore or Near-shore	RO Plant or Direct Drive	2008	Sea Power carrying out ongoing development and tank testing. Currently attempting to reduce LCOE targets.

(Continues)

Table 1.3 (Continued)

Project	Developer	Location	Technology	Site	Distribution	Operation	Description
SDE Sea Wave Power Plant	SDE Energy Ltd.	Israel	Buoy	Nearshore	Hydraulic ram	2010	A breakwater-based wave machine. Device is installed close to shore and utilises vertical pumping motion of buoys for operating hydraulic rams to power generators. First ran from 2008 to 2010, producing 40 kWh at peak.
SeaBased	SeaBased AB.	Sweden	Buoy	Off-shore	Linear generator on seabed	2015	In cooperation with the Swedish Energy Agency, developed its first wave power park on the Swedish West coast. First phase deployed in 2015 and comprising 36 wave energy converters and one substation.
SeaRaser	Alvin Smith Dartmouth Wave Energy-Ecotricity	UK	Buoy	Nearshore	Hydraulic ram	2008	Consisting of piston pumps attached to the sea floor with a float buoy tethered to each piston. Waves cause the floats to rise and fall, pressurising water piped to reservoirs on-shore, driving hydraulic generators. Currently undergoing extensive modelling ahead of a sea trials.
SINN Power Wave Energy Converter	SINN Power GmbH	Germany	Buoy	Nearshore	Linear generator	2014	Wave energy converter installed on Crete in 2016 consisting of a variable number of buoys attached to an inflexible steel frame. Electricity is generated as up-and-down wave motion lifts the buoys, driving rods through a generator unit.

Ocean Wave-Powered Generator	SRI International	USA	Buoy	Off-shore	Electro-active polymer artificial muscle	2004	A wave buoy variant, built using special polymers, being developed by SRI International. Testing of a single wave energy converter module has been ongoing in Crete since 2015. A floating wave energy converter will be deployed in 2018. Market entry with single-module WECs is planned in 2020.
Wavebob	Wavebob	Ireland	Buoy	Off-shore	Direct Drive Power Take off	1999	Wavebob has conducted extensive tank tests and some ocean trials. It is an ocean-going heaving buoy, with a submerged tank which captures additional mass of sea-water for added power and tunability, and has a Tank Venting safety feature.
WaveEL	Waves4 Power	Sweden	Buoy	Off-shore	Hydroelectric turbine	2010	Waves4Power is the developer of buoy based Off-shore Wave Energy Converter. A demonstrator plant was planned for 2015 at the Runde Test Site in Norway. This will be connected via subsea cable to the shore-based power grid.
Wave Piston	WavePiston ApS	Denmark	Oscillating wave surge converter	Nearshore	Pump-to-shore, hydro-electric turbine	2013	WavePiston uses vertical plates to exploit ocean wave horizontal movements.

(Continues)

Table 1.3 (Continued)

Project	Developer	Location	Technology	Site	Distribution	Operation	Description
							By attaching several parallel plates to a single structure the application of 'force cancellation' reduces moorings. Testing and modelling has demonstrated this can reduce mooring structures to 1/10. The structure uses steel wires stretched between two mooring points.
Wave Dragon	Erik Friis-Madsen	Denmark	Overtopping device	Off-shore	Hydroelectric turbine	2003	A strong and flexible structure suited for off-shore use is devised. When the vertical plates move back and forth they pressurise water, which is transported to a turbine through pipes. A central station then converts this to electric power. The Wave Dragon wave energy converter uses large wing reflectors to focus waves up a ramp into an off-shore reservoir. The water returns to the ocean by the force of gravity via hydroelectric generators.
Wave Roller	AW-Energy Oy	Finland	Oscillating wave surge converter	Nearshore	Hydraulic	2019	The Wave Roller is a plate anchored on the sea bottom by its lower part. The waves back and forth surges move the plate. Kinetic energy transferred to this plate is collected by a piston pump. A full-scale Wave Roller 2019 demonstration farm installation was built off Peniche, Portugal in 2019.

Name	Developer	Country			Technology	Year	Description
Wave Star	Wave Star A/S	Denmark	Multi-point absorber	Off-shore	Hydroelectric turbine	2000	Wavestar draws energy from waves using floats that rise and fall with the wave up and down motion. Floats are attached by arms to a platform standing on legs secured to the sea floor. Float motion is transferred via hydraulics into the generator rotation, producing has tested electricity. A 1:10 Wave Star was tested in Nissum Bredning from 2005 and decommissioned in 2011. A 1:2 Wave Star machine was tested in Hanstholm producing electricity to the grid from 2009–2016.
Wave Carpet	Paul Mario Koola	USA	Very Large Flexible Float- ing Structure	Off-shore	Smart Materials	2003	Wave Carpet is a novel deep off-shore wave-power floating system concept funded by the US Navy. It will have a low overall life-cycle cost due to an integrated design, be rapidly re-deployable, easy to maintain and have inherent design reliability, ensuring steady power output from a randomly fluctuating be resource.
Parasitic Power Pack, P3	Paul Mario Koola	USA	Power for 4 inch diameter Sonobuoys	Aircraft Deployed Sensor		2010	A robust maintenance-free military Parasitic Power Pack, P3, modularly inserted into 'free floating' buoy systems

(Continues)

Table 1.3 (*Continued*)

Project	Developer	Location	Technology	Site	Distribution	Operation	Description
							deployed in Distributed Sensor Networks by U S Navy submarines. This increases situational awareness and battlegroup integration by enabling communications at speed and depth.
							It will be dynamically positioned, using built-in energy storage, an internal electric grid, non-corrosive maintenance-free hull design, self-propulsion with advanced controls to minimise tug power. It will act as a wave damper, thereby sharing the cost of power generated.
							P3 will not interfere with the antenna on the upper portion of the buoy and not occupy more than 20 inches in length producing a steady power output of 40 milliwatts with capacity to store 60 Joules of energy.
							P3 will use ocean wave oscillations of the buoy to do this. Unlike regular wave energy devices tuned to ocean waves, this device has a platform whose dimensions are preset for a specific communications purpose.

development is taking place, giving an idea of the scope of technology and its geographical distribution.

The expansion of this list from the previous edition of this book reflects the wide interest in this technology as well as some of the difficulties in developing wave power.

Many devices described from the previous edition have now failed because of the difficulties of the marine environment; however, lessons have been learnt and new approaches adopted.

Falcao (2010) gives an excellent review of these developments.

Table 1.3 shows that development is ongoing for the following prototype categories, in the terms described above in order of the number of prototypes:

- 11 Point absorbers or buoys
- Three surface following attenuators
- Two oscillating wave surge converters
- One oscillating water column
- One overtopping device
- One multi-point absorber
- One underwater attenuator
- One fully submerged wave termination device.

Table 1.3 illustrates the very significant number of wave energy device prototypes being considered for development, with very widely differing architectures.

Relatively few of these projects have been fully installed and tested; there is still a lack of a clear consensus on which are the most applicable and reliable technologies for wave energy extraction.

It is to be hoped that the proposed variety of devices will shake down as an understanding of their technological and operational strengths and weaknesses increases.

Four technologies recently developed, installed and operational are shown in Figure 1.4, building an increasing wave power knowledge base.

1.7   MEC device learning or experience curves

In the light of the above review, an important task for marine renewables is to learn technologies and methodologies from other power industries, and this has been described in some detail for on-shore wind in Mukora (2013).

Learning curves were defined by Neij (1999) as the cost reduction of a standardised product from a single firm, whereas experience curves are defined as the cost reductions of non-standardised products on a national or global level.

Neij identified that the percentage fall in costs for a doubling in cumulative installed capacity can be defined as the learning rate for a product. The progress ratio is (1-learning rate), for example a progress rate of 95% means that only a 5% improvement has been made over a doubling of capacity, perhaps it could better be described this as a 'low progress rate'.

An example of technology experience curves for other power generation industries, photovoltaic, wind and gas turbines is shown in Figure 1.4. This gives

Figure 1.4   *Four wave device technologies tested in last ten years, taken from Table 1.3. (a) LIMPET, (b) Wave Dragon, (c) Oyster, and (d) Pelamis.*

the measured energy costs in $/kWh from various newly installed technologies against their installed capacity in MW, both axes plotted logarithmically.

Each technology line shows the dates over which these costs fall and represent their technology experience curves. For photovoltaics, US windmills and early US gas turbines, these represent learning rates of about 20% or progress ratios of 80%.

Because of the marine environment, the lack of consensus on the most applicable extraction technology and a relatively small installed base in MW, it is likely that MEC learning rates will be lower than for wind or gas turbines, as suggested in the graph.

Some authors, including Prof Salter, cautioned that:

'Learning curves cannot be used to assess Wave & Tidal energy as there is currently little or no market experience and hence no data '. . .' to apply such an approach for the immature ocean energy industry, with varied technologies, could be little more than guesswork'.

However, this author suggests that engineers need some operational experience and guidance to lay the ground-work for improvement and that some learning data is now becoming available. This is what this book is about.

That caveat must be applied to the MEC's line in Figure 1.5.

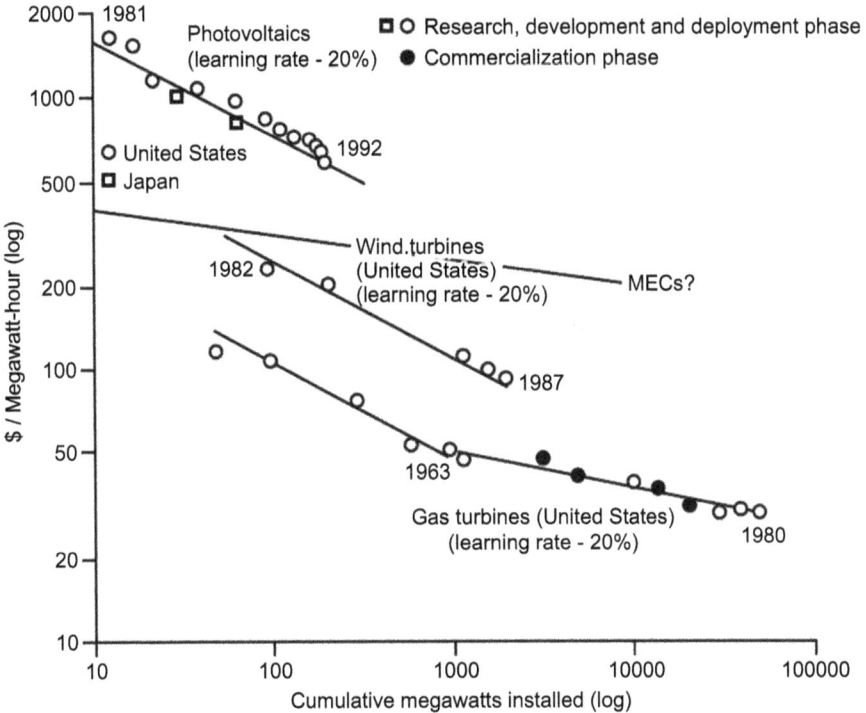

Figure 1.5   Japan and US experience curves for photovoltaics, wind turbines and gas turbines, Mukora (2013)

1.8   Current world wave and tidal developments

1.8.1   Europe

European wave developments have been largely concentrated in northern European countries because of their exposure to Atlantic and North Sea wave energy, see Figure 1.1, particularly in Ireland, the United Kingdom, Iceland, Norway, Denmark, Germany and the Netherlands.

There are research activities in all these countries, some of which are associated with European Union research projects.

Cornwall, Western Ireland, Scotland and Wales all have a number of interesting projects exposed as they are to the Atlantic resource with potentials up to 70 kW/m.

Tidal developments in Europe are currently mainly concentrated around France, the United Kingdom and Ireland, see Figure 1.2, where the tidal range from the Atlantic is amplified by interaction with the entrance to the Manche, the English and St George's Channels and the large bays in the French, Irish and British coasts. The Rance Tidal Power Station near St Malo with a peak power output of 240 MW is the world's longest-standing tidal power barrage installation.

One of the most exciting developments in the United Kingdom had been the reconsideration of the Severn Estuary tidal resource, because of its potential, by the possible exploitation of the Swansea Bay tidal barrage, intended to operate a barrage lagoon on the incoming and outgoing tides using high efficiency, through-flow, low-head turbines with a peak power output of 320 MW.

1.8.2   The Americas

Wave developments in North America are concentrated on the Pacific seaboard in the northerly states of the United States, i.e. Alaska, Washington and Oregon with wave power potentials of 50–60 kW/m, and on the Atlantic seaboard in the state of Maine and in Canada in the provinces of Nova Scotia, Newfoundland and Labrador with lower wave power potentials up to 50 kW/m, see Figure 1.1.

In South America, wave power interest is primarily on the Pacific seaboard in the southern Chilean regions.

Tidal developments in North America are concentrated on the Pacific seaboard in the northerly states of the United States, i.e. Alaska, Washington and Oregon and in Canada in the province of British Columbia around the Bay of Fundy, where the tidal range from the Atlantic is amplified by bay geography, see Figure 1.2.

1.8.3   Asia

There are a few wave developments currently in Asia.

But tidal developments are being considered on the eastern seaboard of India where the coastal orography in the Rann of Kutch concentrates the tides.

There is also considerable interest in tidal power in the Bo Hai Sea between China and the Korean peninsula where the geography assists the tidal range. Korea already has a significant barrage installation on its east coast at Sihwa Lake, which at peak power of 254 MW is currently the world's largest tidal power station, Figure 1.2.

Japan also has modest wave resources but is considering tidal schemes where there are current concentrations between the islands in the archipelago.

1.8.4   Oceania

Both Australia and New Zealand have been considering wave and tidal schemes.

In Australia, wave resources on the South-East coast of Western Australia are promising up to 70 kW/m.

Tidal schemes are being considered on the Northern coast of the Northern Territories and the North East coast of Queensland, whilst locations around the Bass Strait between Victoria and Tasmania have potential.

New Zealand has strong wave resource, particularly on the West coast of South Island and tidal potential in the Cook Strait between North and South Island.

1.9   Wave and tidal power economic terminology

1.9.1   Terminology

The definition of availability for MECs has not yet been established. However, since 2007, an International Electrotechnical Commission working group has been working to produce a standard IEC 61400-Pt 26 to define WT availability in terms of time and energy output.

When that standard is published, there will be an internationally agreed definition of WT availability in terms of time and energy which could also be applied to MECs. However, two availability definitions have been generally adopted in the United Kingdom for WTs applicable to MECs and are summarised below.

- Technical availability, also known as system availability, is the percentage of time that an individual MEC or MEC farm is available to generate electricity expressed as a percentage of the theoretical maximum.
- Commercial availability, also known as intrinsic availability, is the focus of commercial contracts between MEC farm owners and MEC OEMs to assess the operational performance of a MEC farm project. Some commercial contracts may exclude downtime for agreed items, such as requested stops, scheduled repair time, grid faults and severe weather, when MECs cannot operate normally.

For the rest of the book, the term availability refers either to technical availability as defined above lending itself to comparison from project to project.

From the above definitions, it follows that technical availability will always be lower than the commercial availability because there is more alleviation of downtime for the former and an important issue off-shore is that availability, $A(u,t)$, is affected by both time and wind speed, u.

In respect of reliability the following expressions are useful:

$$\text{Mean Time to Failure, MTTF} \tag{1.1}$$

$$\text{Mean Time to Repair, MTTR} \tag{1.2}$$

$$\text{Logistic Delay Time, LDT} \tag{1.3}$$

$$\text{Downtime, } MTTR + LDT \tag{1.4}$$

$$MTBF = MTTF + MTTR + LDT \tag{1.5}$$

$$MTBF \approx MTTF \tag{1.6}$$

$$MTBF \approx MTTF + MTTR = 1/\lambda + 1/\mu \tag{1.7}$$

$$\text{Failure rate, } \lambda\lambda = 1/MTBF \tag{1.8}$$

$$\text{Repairrate, } \mu\mu = 1/MTTR \tag{1.9}$$

Commercial availability,

$$A = (MTBF - MTTR)/MTBF = 1 - (\lambda/\mu) \tag{1.10}$$

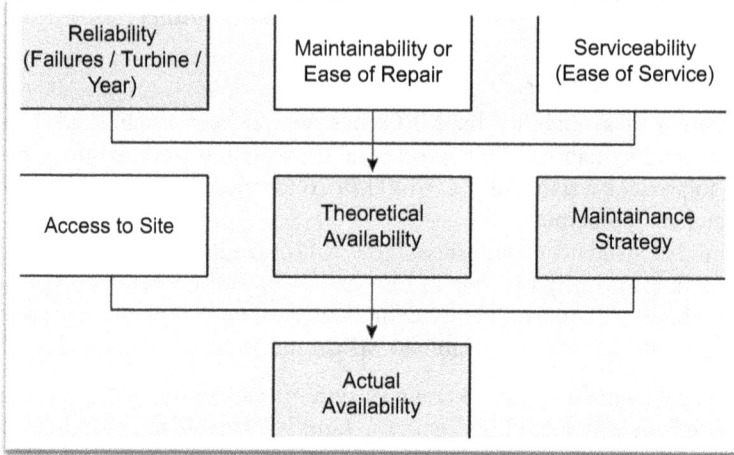

Figure 1.6   Availability as a function of MEC properties, access to site accessibility and maintenance strategy

Technical availability,

$$A = MTTF/MTBF < 1 - (\lambda/\mu) \tag{1.11}$$

Note that these are all expressed in terms of the variable time but availability can be expressed in terms of energy production and this will ultimately be more valuable for the operator than a definition involving reliability. Figure 1.6 shows the linkage between availability and reliability.

Capacity factor and specific energy yield are two terms that can be used to describe the productivity of a MEC or MEC farm as they are for WTs. Capacity factor, C, is defined as the percentage of the actual annual energy production E (MWh) over the rated annual energy production, *AEP,* from a MEC or MEC farm of rated power output P:

$$C = AEP \times 100 /(P \times 8760)\% \tag{1.12}$$

Specific energy yield, S (MWh/m^2/year) is defined as the *AEP* of a MEC normalised to the swept area A (m^2) of the device:

$$S = AEP/A \tag{1.13}$$

The ratio R_S of rated power, P, over the swept area, A, is a fixed value for a specific MEC type:

$$R_S = P/A \tag{1.14}$$

Or:

$$R_S = S/(C \times 8760) \tag{1.15}$$

For a specific type of MEC, the specific energy yield is proportional to the capacity factor:

$$S = R_S \times C \times 8760 \tag{1.16}$$

Therefore, the operational performance of a MEC or MEC farm can be defined as the percentage of the achieved over the expected C or S.

1.9.2 Cost of installation

Marine power uses large MECs whose capital cost is currently estimated at around £4–8M/MW, compared to on-shore WTs at £0.85M/MW; however, MECs are in an early stage of development and on-shore WTS are fully developed, see Figure 1.5. MEC structures are large, initially structures that will be installed in relatively shallow water depth, 5–20 m, and the weight of each structure will be relatively low, ≈ 400 tonnes, depending on rating. So, in contrast to typical oil and gas off-shore structures, the applied vertical load to the foundation is relatively small but the overturning moments of tidal and wave devices are large. Therefore, a MEC foundation or mooring may account for larger proportion of the installed cost. So MEC unit capital costs are large and will increase as the MEC farms are placed in deeper water.

However, a single MEC design can be mass-produced for use over a whole MEC farm or many MEC farms, rather than each structure/foundation being individually engineered, as it would be in the oil and gas industry. So MEC capital costs will fall progressively with subsequent projects at later times, which is gradually being noted in Danish, Swedish, UK, German and Dutch off-shore wind projects.

An interesting comparison can be made in Figure 1.7 between the weight-to-power ratio, installation capital cost and cost of energy for MECs compared to off-shore WTs; however, the MEC costs must be treated with care as the data comes from relatively few projects, but Figure 1.7 clearly shows the increased weight/power ratio of MECs and their consequent higher installation and energy costs.

1.9.3 Cost of energy

Cost of energy (*CoE*) will be used to evaluate the economic performance of different MEC farms. This methodology was adopted in a joint report by the International Energy Agency (IEA), the European Organisation for Economic Co-operation and Development (OECD) and US Nuclear Energy Agency (NEA). It compared the cost of different electricity production options. A simplified calculation equation was adopted in the United States to calculate the *CoE* (£/MWh) for a MEC system:

$$CoE = (ICC + FCRO\&M)/AEP \tag{1.17}$$

where:
ICC is initial capital cost (£); *FCR* is annual fixed charge rate (%);
AEP is annual energy production (MWh); *O & M* is annual O&M cost (£).

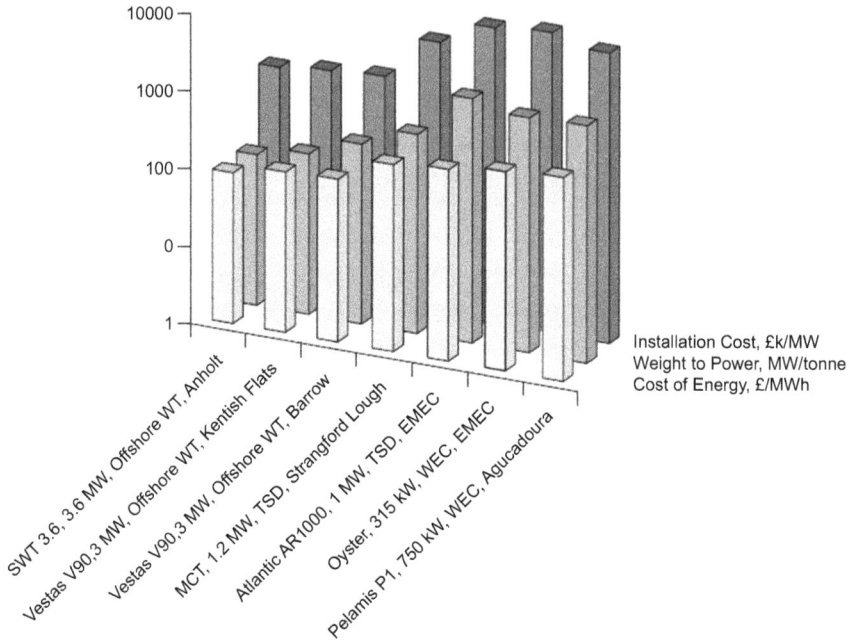

Figure 1.7   Comparison of off-shore WTs and MECs by installation cost, power weight and cost of energy

The result of this approach is the same as that of levelised electricity generation cost, where the parameter FCR is a function of the discount rate r used, as follows:

$$FCR = r/(1 - (1 + r)^n) \tag{1.18}$$

where $r \neq 0$.

The discount rate r is the sum of inflation and real interest rates. If inflation is ignored, the discount rate equals the interest rate. For the special case of a discount rate $r = 0$, unlikely in the real world, *FCR* will be *ICC* divided by the economic lifetime of the MEC farm in years, currently estimated at $n = 20$ years.

Early studies with off-shore wind show clearly that operators who impose a higher quality O&M regime achieve higher availability, lower through-life costs and a lower CoE. The relationship between CoE and the design and operations of the MEC is shown in Figure 1.8, and the focus of this book is on the highlighted areas of the diagram.

1.9.4   O&M costs

The estimated cost of MEC farm energy will vary depending on the site, project and MEC design, but it is clear that MEC projects will be significantly more costly than off-shore wind, at least in the first instance.

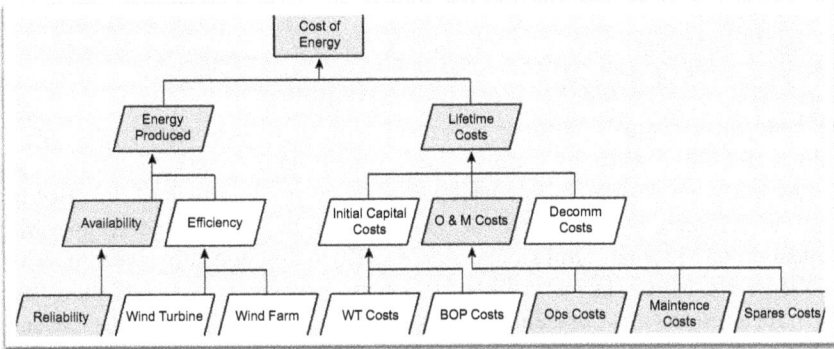

Figure 1.8   Structure of cost of energy, showing highlighted in grey areas of interest for this book

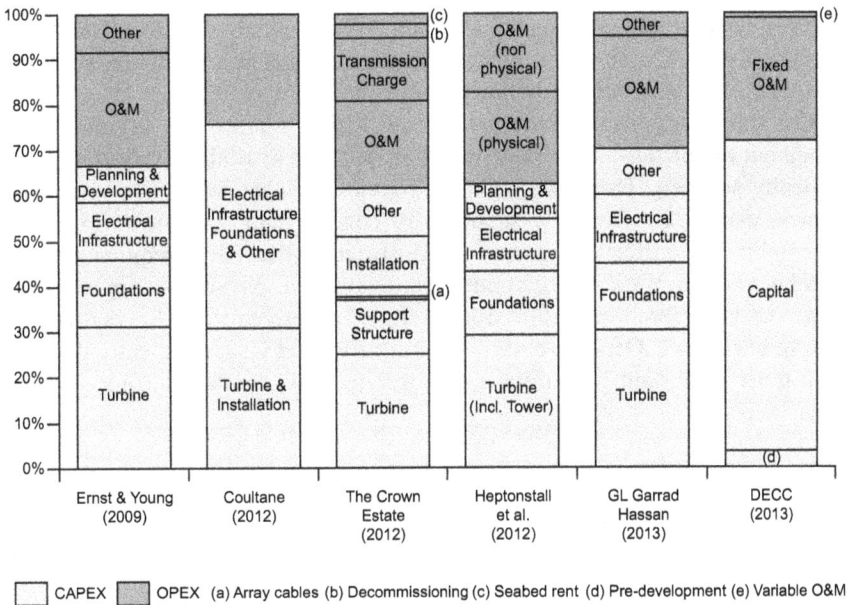

☐ CAPEX ■ OPEX (a) Array cables (b) Decommissioning (c) Seabed rent (d) Pre-development (e) Variable O&M

Figure 1.9   Comparison of the share of CAPEX and OPEX costs for off-shore wind farms from various sources

As MEC designs become adapted to off-shore conditions, the achievement of a favourable economic solution depends upon controlling the MEC farm system full life-cycle cost.

Figure 1.9 shows a comparison between estimated CAPEX and OPEX costs for off-shore wind, taken from Crabtree *et al.* (2015), where:

(a) Array cables;
(b) Decommissioning;

(c) Seabed rent;
(d) Pre-development;
(e) Variable O&M.

The object for MC farm development will be to concentrate on reducing CAPEX and OPEX costs by learning the lessons from off-shore wind power, perhaps by using the design assets developed in that industry. Much of the price premium to be paid by ME farm developers will be attributable to the MEC Foundation or Mooring, Grid Connection and Operation & Maintenance (O&M).

O&M for MECs farms is likely to be more complex than off-shore wind farms. As a result, O&M percentage costs are likely to be higher than the measured 23%–36% shown in Figure 1.9.

Off-shore conditions for MECs require more onerous erection and commissioning operations; meanwhile, accessibility for off-shore routine servicing and maintenance is a major issue.

During winter, a whole MEC farm may be inaccessible for many days due to harsh sea, wind or visibility conditions. Even given favourable weather, O&M tasks are more costly than on-shore, being influenced by distance off-shore, site exposure, MEC farm size, MEC reliability and maintenance strategy.

Off-shore conditions require special lifting equipment to install and change out major sub-assemblies, which may not be available at short notice or be locally sourced. Therefore, advanced techniques are needed to plan maintenance, using data from the supervisory control data acquisition (SCADA) and condition monitoring systems (CMS) fitted to the MEC, requiring a thorough knowledge of off-shore conditions, qualitative physics theory and other design tools to predict failure modes in less conventional ways than has hitherto been done. Off-shore remote monitoring and visual inspection become much more important to maintain appropriate MEC availability and capacity factor levels.

1.9.5   Effect of reliability, availability and maintenance on cost of energy

Equation 1.13 for *CoE* can be expressed as a function of λ and μ allowing us to see the effect of reliability and maintenance on *A* and *CoE* as follows:

$$CoE = (ICC \times FCR + O\&M(\lambda,\ 1/\mu))/AEP(A(1/\lambda,\mu) \tag{1.19}$$

Reductions in failure rate λ, will improve reliability *MTBF*, $1/\lambda$, and availability, *A*, therefore reducing O&M costs. Reductions in downtime *MTTR*, will improve maintainability, μ, and availability, *A*, therefore also reducing O&M. As a consequence *CoE* will also reduce as λ and μ improve.

Two useful reports from Renewables UK (2013) and Strategic Initiative for Ocean Energy (2014) summarise current UK/European economic views of wave and tidal energy, demonstrating the reliability, availability and maintenance steps needed to improve viability.

These reports indicate, respectively, energy costs of 25–50 £/MWh & 25–65 Euro/MWh for both wave and tidal energy MECs, which looks ambitious compared to those indicated by Figure 1.5 for this stage of the learning curve.

1.10 Roles

1.10.1 General

There are many stakeholders involved in inventing, developing, building and operating MECs and their farms; their actions will define and shape our ability to achieve the objectives of those farms and their interrelationship with capital is shown in Figure 1.10.

Those objectives are to generate electricity reliably from the sea at competitive prices, and to provide an acceptable return to each stakeholder. The following describes the role of each of these major stakeholders, so that the reader can understand their influence upon innovation, development, building and operation.

1.10.2 Innovators

The technology of MECs, WECs and TSDs in particular is still in its infancy and there is a need for innovation. Innovators are, therefore, very present in the marine energy field and their input is essential to the healthy development of reliable plants. There is currently a wide range of wave and tidal devices under development, as shown in previous sections, and it would be impossible to identify the reliability characteristics of all of them. However, some principles are beginning to emerge.

1.10.3 Governments

Governments set national energy policies, which have played a large part in stimulating the renewable energy industry. The motivations of these policies are to

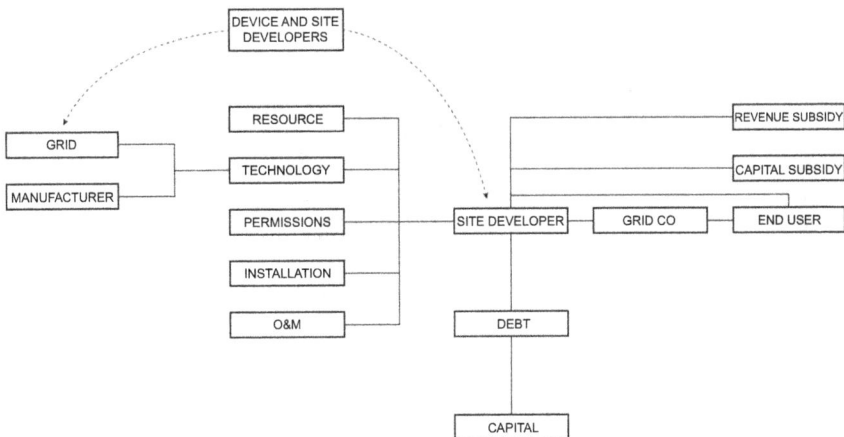

Figure 1.10 Relationship between roles and capital in the development of MEC farms, Hardisty (2009)

secure long-term energy resources for the country, reduce greenhouse gas emissions, stimulate the economy and answer democratic calls for more sustainable energy production. These energy policies are empowered by subsidies for certain energy sources which are highly contentious in democratic countries, where citizens demand to see personal benefits in energy costs from these subsidies. These issues are complex and vary from country to country but include capital subsidies to offset the cost of construction and revenue subsidies as feed-in payments. The former stimulate the MEC construction process, the latter provide long-term security for the MEC operator.

However, the European Union (EU) has so-far sustained a degree of cohesion between member countries with respect to subsidising renewable energy sources. This has led to the EU becoming a world leader in their development, manufacture and deployment.

It is, however, a truism that all electrical generation sources, since their inception in the 1880s, have received some sort of subsidy to sustain early technological development. This was especially true for large-scale deployment of new technologies, including the use of particular fuels, such as coal and nuclear or the application of high-speed steam or gas turbines. This has been so whether the subsidies came from governments, innovators, developers, original equipment manufacturers (OEM) or the consumers themselves via elevated electricity costs.

Our current array of reliable and profitable electrical energy supply sources, developed over the last 130 years, is a tribute to the efficacy of that process. In fact, the United Kingdom is a very good example of the early development of a national electricity network, provided by the national grid, incorporating large coal-fired steam turbine generators entirely capitalised by the government and subsidised from elevated electricity charges, which ultimately fell to acceptable low levels.

1.10.4   Regulators

In the United Kingdom, the regulator, the Office of the Gas and Electricity Markets (OFGEM), is setting up the market for marine energy extraction. A particularly important aspect of this work has been the development of a role for off-shore transmission owners (OFTO), who ensure that marine energy extraction farms have a secure and flexible connection asset, to transport energy from off-shore into the on-shore transmission grid. The long-term availability and reliability of the OFTO's connection asset is essential to the achievement of a marine energy extraction farm's objectives. Its technical reliability, however, will be outside the scope of this book. The regulator implements the subsidies, enacted by government, that are available to renewable energy generators.

1.10.5   Investors

Investors in marine energy extraction include banks, energy supply and energy manufacturing companies and landowners, including, in the United Kingdom, the Crown Estate and Crown Estate (Scotland), who have licensed the off-shore areas for development. The issues of reliability and availability of the MEC farm asset are of most importance to the long-term investors, and the energy supply and manufacturing companies, since this is the means by which their investment can be reliably and predictably repaid, with the required return. The difficulty for investors, in this emerging technology, is to understand the technical issues involved so that the right parameters can be defined for their investment. It is also true that investors are keen to benefit from short-term government subsidies for renewable energy, rather than for the profits available from the long-term operation of reliable renewable energy sources. By explaining the technical issues of MEC farm reliability and availability, this book will enable those parameters to be more precisely defined.

1.10.6 Certifiers and insurers

Certifiers, such as Germanischer Lloyd and Det Norsk Veritas, will be responsible for ensuring that MEC designs and their associated marine structures meet IEC standards and are therefore insurable. Project insurers are also important participants as they determine the premiums for large off-shore projects, which affect their investability. An important aspect of these processes is imposing the necessary Health and Safety (H&S) regime on the installation and operational phases of the project to ensure that the human risks are acceptable and that the losses due to such risks are minimised. These processes were developed for the on-shore wind industry and have successfully ensured that machines and structures are sound and safe investments. They are even more important off-shore, where the environment is more challenging. However, this has meant that, so far, MEC designs may have focused more on meeting safety and certification requirements than on production demands. That will change as investors demand a return on their investment during operation.

1.10.7 Developers

Developers of marine energy extraction farms are emerging as consortia of investors, energy companies, MEC manufacturers and operators. Their objective is to gain a financial return when the assets are subsequently sold on to long-term operators, such as the main electricity generating companies. Because of the scale and complexity of the off-shore asset, these consortia are attempting to attract long-term Investors as part of the development team, and that requires financial experts to have a better understanding of the technical issues concerned. Hardisty (2009) has coined the term device and site developers (DSD) for these types of consortia.

A major part of the deployment of marine energy extraction farms depends upon the marine installation assets, which include port and docking facilities, installation vessels, maintenance vessels fleet and the manpower and infrastructure to manage and operate these assets. The latter are usually provided by civil and marine engineering businesses who are starting to become important members of MEC farm developer consortia.

1.10.8   Original equipment manufacturers

The principal OEMs involved are MEC OEMs, some of whom are owned by larger energy groups, such as Siemens, with large balance sheets and the potential to invest in promising new technology. However, many of the MEC OEMs are small engineering companies, often start-ups, incorporating the original device innovators, therefore their capital base is small and they rely on larger investors and support from governments. But a MEC farm will be a complex generation, collection and transmission asset, with a substantial balance of plant (BOP), including transformers, cables, switchgear and built assets. This requires cable and transmission OEMs as well, so any large MEC farm will involve a consortium with a number of different OEMs, of whom those with the largest capital will take the majority of the risk.

The regulator's actions are encouraging transmission OEMs to participate in the off-shore cable OFTO activity, to mitigate the off-shore risks involved. However, conventional on-shore transmission OEMS will still have a significant financial, management and technical role in the collection and off-shore substations of an MEC farm.

1.10.9   Test facilities

Reliable devices for marine renewable energy can only be developed if there is a substantial investment in prototype testing, both of devices and their components.

To facilitate this, there has been considerable development of National and University testing facilities, which have been prominent in the United Kingdom, particularly in Scotland, where there are particularly strong wave and tidal energy sources and the Scottish government has encouraged industry and universities to engage in the development.

The aim has been to prove the viability of scaled devices and then to prove full-size prototypes in full-scale tidal and wave conditions.

Such facilities are essential to the future development of marine renewable energy.

The world's largest and longest operating off-shore renewable test facilities are in the United Kingdom, at the European Marine Energy Centre, EMEC in Orkney, Scotland, led by its Managing Director, Neil Kermode OBE.

EMEC has the following facilities, see Figure 1.11:

• A headquarters and deployment facility;
• A wave test facility at Billia Croo;
• A tidal test site at the Fall of Warness.

1.10.10   Operators and asset managers

The operators of marine energy extraction farms will be large energy companies providing electricity into the transmission grid. Most of these operators are broad-technology generators with fossil- and nuclear-fired and renewable generation

(a)

(b)

(c)

Figure 1.11 *European Marine Energy Centre in Orkney, Scotland, UK, courtesy EMEC. (a) EMEC headquarters and deployment facilities; (b) Billia Croo wave facility, Wello device being installed; and (c) Fall of Warness tidal facility, turbine being installed.*

assets. In view of the technical complexity of MEC assets, a few specialised off-shore operators are developing, particularly in the Scandinavian market. These operators are building their expertise to match their existing assets in on-shore wind, hydro and gas-fired generation.

It seems likely that, in the future, additional specialised operators will come into the market, but the size and complexity of marine energy extraction assets mean that these will be large companies with diverse international portfolios of assets, developed to balance their exposure and risk in this sector. As the industry matures the current certification- and safety-oriented approach is likely to change: more stringent demands for return on the larger capital outlays will encourage a more vigorous production-oriented approach. In this stage of industry development, the interaction between operators, asset managers, certifiers, insurers and investors will be strengthened.

1.10.11 Maintainers

Maintainers work for a variety of the MEC farm stakeholders.

MEC OEMs, as yet, have not developed service departments, but they have access to the signal conditioning and data acquisition (SCADA) data streaming from their machines during commissioning and warranty periods. They also have detailed knowledge of the development of their own MECs through prototype tests, supply chain development and production tests. Their staff are trained on their machines and have built up a detailed personal knowledge of the idiosyncrasies of individual MEC types. This expertise is deployed during the warranty period, regulated by the project contract. For some MEC OEMs, this may change with time as they recognise the benefit to their business of the O&M market, and the importance to the developers and operators of through-life performance.

Operators will gradually gain experience of MEC farm operation, different in nature to that of the MEC OEM, being more focused on production needs and the through-life performance of the asset. They will have their own management and some of their own O&M staff, but may rely upon sub-contractors and the MEC OEM for some of that support. However, they frequently lack detailed knowledge of individual MEC farm equipment and rely, in large part, upon the warranty period to gain that knowledge and experience.

Operators may opt to continue with a maintenance contract with the MEC OEM after the completion of the warranty period, but marine energy extraction farm operators are likely to be large; with experience on other MEC farms, many will opt to undertake their own O&M. This will allow them to impose their own asset management objectives upon the MEC farm and ensure long life.

MEC farm maintenance will rely heavily upon the expertise of the management and staff carrying out this highly skilled activity. MEC farm design, choice of MEC, availability of appropriate access assets, spares and tools can facilitate the activity, but success is impossible without staff who are well trained in H&S and the technology of the asset. This is an important issue that will be addressed later.

1.11   Summary

WEC and TSD, or generically MEC, are emerging technologies that extract energy from an intermittent renewable resource of potentially higher density than wind power: although exposed to that resource 24/7, they do not operate at full power most of the time. They must, therefore, be designed to withstand all the loads the resource imposes, under all these intermittent operational conditions. Permanently installed at selected wave or tidal sites, they will endure harsh climatic, current and wave load conditions, which their complex mechanical, electrical and control systems must withstand. Therefore, their reliability and survivability will be an engineering challenge. Although this could be mitigated by selecting benign wave or tidal sites, this could reduce their energy effectiveness.

This chapter has outlined the wave and tidal resource available to these devices, which is substantial. However, the location for resource extraction, particularly in the case of WECs, and less so in the case of TSDs, imposes difficult challenges for those attempting to design reliable devices.

Chapter 2

Resource

2.1 Introduction

Power generation from the sea can exploit either sea waves or tidal flows.

The main attraction of this form of generation is the combination of a substantial resource and relatively predictable output, although less so in the case of waves.

Against this are the challenges of constructing reliable systems that can generate economically in a hostile environment.

Furthermore, while tidal power at sea is repeatable and predictable, wave power is much less repeatable or predictable.

This marine energy field is immature and, unlike wind generation, there is no obvious preferred solution.

Interest in generating power from the sea rose sharply after the first 'oil shock' in the 1970s, but spending on development fell with reducing oil prices and a clearer understanding of the barriers to implementation. However, recent, and growing, concerns about carbon dioxide emissions, combined with perceived difficulties in further wind power developments, have led to a renewed interest in the subject. There is much activity, supported by both public and private money, in searching for practical and economical methods of generation.

The nature of the resource depends on the generation perspective. As has been identified, marine generation devices can either exploit tidal flows or extract energy from sea waves. Tidal flows can be in and out of an estuary, the barrage on the River Rance being an example, or a body of water connected to the sea, Strangford Lough being a dramatic example. They can also be ocean currents, perhaps flows around an island such as the Alderney Race around the eponymous Channel Island. Waves can be exploited at the shoreline, as in the Limpet device, near shore in shallow water or in deeper water.

The choice of operating conditions determines the choice of marine energy device, and that is in turn influenced by the extent of exploitable resource. The availability of a resource is a sine qua non for its exploitation but there are many practicalities to be considered. These include the availability of suitable marine energy equipment to exploit the resource, the survivability of that equipment in the proposed environment, issues to do with shipping, environmental and amenity issues and the need for a connection back to the shore.

Many wave energy systems have been demonstrated and even more have been proposed. They can be classified in several ways, first by intended position.

- Shoreline devices. These are fixed rigidly to the shore.
- Near shore devices. These are tightly moored to the seabed, close to the shore, in depths of water from 10 to 25 m.
- Deep-water devices. These are not in direct contact with the seabed – mooring is used to maintain general position.

Systems can also be classified by the characteristics of the energy absorption mechanism exploited.

- Point absorbers. These move with the incident wave and are small compared to the wavelength of the wave.
- Terminators. These have their principal axis parallel to the incident wave front.
- Attenuators. These have their principal axis perpendicular to the incident wave front.

Power can be extracted from tidal streams by turbines and a range of designs exist. Some are open turbines similar to wind turbines (WTs); others have gone for ducted designs. Various kinds of oscillating flaps have also been tested.

2.2   Potential sea wave resource

2.2.1   *Geographical distribution of resource*

Oceans cover approximately 70% of the area of the planet. Sea waves are generated by wind blowing over the sea surface, with high wind speeds correlating with high sea waves. The waves contain a large amount of energy that travels thousands of kilometres before reaching a shoreline. Countries that have significant lengths of coastline are obviously best placed to exploit wave generation and the United Kingdom is particularly fortunate in this respect. Japan, the Philippines, China and Indonesia are amongst other countries with good resources. Estimates of the available wave power resource have been made by various authors and agencies (Mørk *et al.* 2010). This is normally expressed as the power per unit length of wave front and Figure 2.1 shows resource on a global scale.

Whilst there are several regions with high values, north-west Europe experiences particularly strong waves from the Atlantic breakers. A closer view of Europe is shown in Figure 2.2.

The British Isles are particularly well favoured and the potentially available resource is shown in Table 2.1. However, not all the resource can be exploited for a range of reasons including engineering difficulties, shipping and environmental reasons, as noted earlier. The shoreline resource is the easiest to exploit but the wave intensity is lower and the acceptability of shoreline installations is low. The available resource rises as the depth of water increases but engineering practicalities begin to dominate, as well as shipping issues.

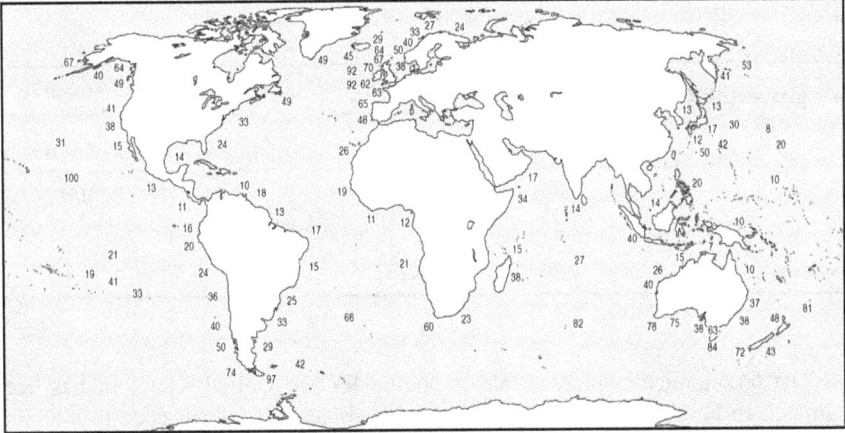

Figure 2.1   World-scale wave power resource close to the shore. Units are kW/m of wave front.

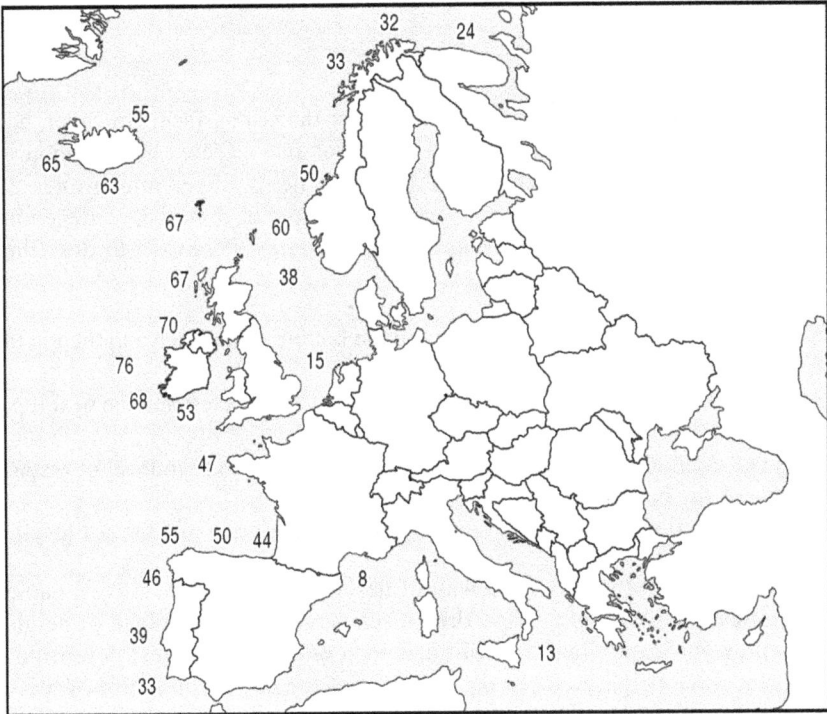

Figure 2.2   Western European wave power resource. Units are kW/m of wave front.

Table 2.1   Economically recoverable wave resource for the United Kingdom

To water depth (m)	Average available		Average recoverable	
	GW	TWh	GW	TWh
100	80	700	87	100
40	45	394	10	87
20	36	315	7	61
Shoreline	<30	<262	0.2	1.75

The 'economically recoverable' resource for the United Kingdom has been estimated to be about 90 TWh per annum or about 25% of the country's annual consumption of electricity, as shown in Table 2.1. 'Economically recoverable' implies that the cost of energy less subsidy is acceptable. The level of subsidy for marine energy is currently high, reflecting the risk and immaturity of the industry. The current level is 3 Renewables Obligation Certificates (ROCs), but this scheme is to be replaced by a 'contracts for differences' scheme and the use of a Current Strike Price.

2.2.2   Wave characteristics

Ocean waves are formed by wind blowing over the sea surface. However, each wave is not only created by local wind speed but also affected by neighbouring waves and small ripples. Because of the existence of upcoming waves and small ripples, some new-born waves can be enlarged or diminished, depending on their relative travelling direction. The usual parameters used to describe a wave are:

- Wave height: vertical distance between the bottom of a trough and the top of a nearby crest.
- Wave length: horizontal distance between one top of a crest and the next top of a crest.
- Wave amplitude: distance between the top of a crest and sea level or distance between the bottom of a trough and sea level.
- Wave elevation: distance between any point of the wave and the sea level.

There is another widely used way of describing a series of waves, namely the significant wave height. Wind-driven sea waves exhibit a range of heights, some with significantly larger amplitudes than others. A practical definition of significant wave height is the average height of the largest one-third of waves, in a measured time period. For instance, ocean waves might be measured in a few minutes and 120 wave crests picked up. In these wave crests, the 40 largest waves are picked to calculate the average height, which is known as the significant wave height.

There are several ways to classify sea waves, but the most well-known method is to categorise the waves based on their directional spreading as listed below:

- Long-crested waves: waves travel in the same direction and have long crests.
- Short-crested waves: waves travel in different directions and have relatively short crests.

Long-crested waves are of interest for wave power generation.

A mathematical description of sea waves is given by Cruz (2010). The available energy can be calculated from characteristics of a wave, as shown in Figure 2.3. The wave shown is a pure sine wave of single period also known as a monochromatic wave; real sea waves exhibit a spectrum.

The wave has amplitude η, height H and wavelength λ, although the period of the wave, T, is more useful. Therefore, the mean power per unit width of wave front, P, is given by:

$$P = \rho g^2 T H^2 / 32\pi \qquad (2.1)$$

where ρ is the density of sea-water and g is gravity acceleration. The available power depends, therefore, on the square of the wave height. An example of two different wave conditions is given in Table 2.2.

Larger amplitude waves have a higher energy density and more destructive power. Real sea waves, however, have a more complex behaviour than this example, with a spectrum of wavelengths. These points are discussed in the next section.

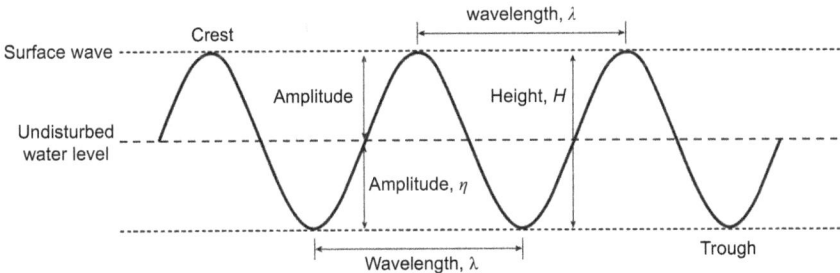

Figure 2.3   Ideal single period, monochromatic sea wave

Table 2.2   Available power at different wave heights

Wave height, H (m)	Period, T (s)	Power, P (kW/m)
0.75	6.5	3.6
5.0	12.0	300.0

2.2.3   Time- and frequency-domain definition of waves

Waves can be considered in either the time or frequency domain.

2.2.3.1   Waves in the time domain

In the time domain, a series of wave elevations with respect to time are used to describe a wave during the chosen time period. The time-domain wave signal can be expressed as a continuous wave or a discrete wave. The continuous wave is used to describe a function which contains no gaps between all of the values. Figure 2.4 shows a random wave in both continuous and discrete form. In this figure, T_e and T_k represent two random complete wave periods. In this context, the wave period is defined as the time interval between two successive zero up-crossings. In other cases, a wave period can also be defined as the time between two successive zero down-crossings or the time distance between two successive peaks.

The first two definitions are suitable for random waves, because several peaks can occur during a time period but without any zero crossing. For instance, in the period T_k in Figure 2.4, there are two positive peaks but only one zero crossing. A wave height is the distance between a positive peak and its adjacent negative peak as indicated by H in the figure. What should be noticed is that the measurement needs to be considered in one complete wave cycle. Here, one wave period is the time between the start of one zero up-crossing to the next zero up-crossing.

2.2.3.2   Waves in the frequency domain

A complex wave can be decomposed into different single-frequency waves. Each individual wave has unique phase information that can be applied to recombine frequency components to generate the original time-domain signal. To demonstrate, a time-domain wave which consists of three sinusoidal waves with different amplitudes, frequencies and phases is shown in Figure 2.5.

If the frequency information of a signal is required, it can be obtained by applying the Fourier transform. Figure 2.5 is the frequency-domain representation of a time-domain wave shown in Figure 2.4. It is shown that the integrated signal is composed of four different sinusoids with angular frequencies ω_1, ω_2, ω_3 and ω_4 rad/s. Each amplitude shown represents its corresponding amplitude in the time

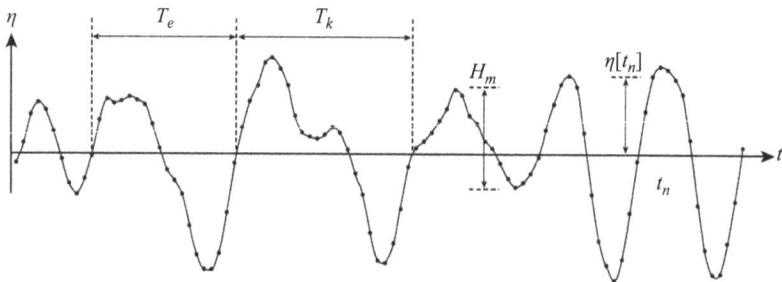

Figure 2.4   Important parameters of a random wave

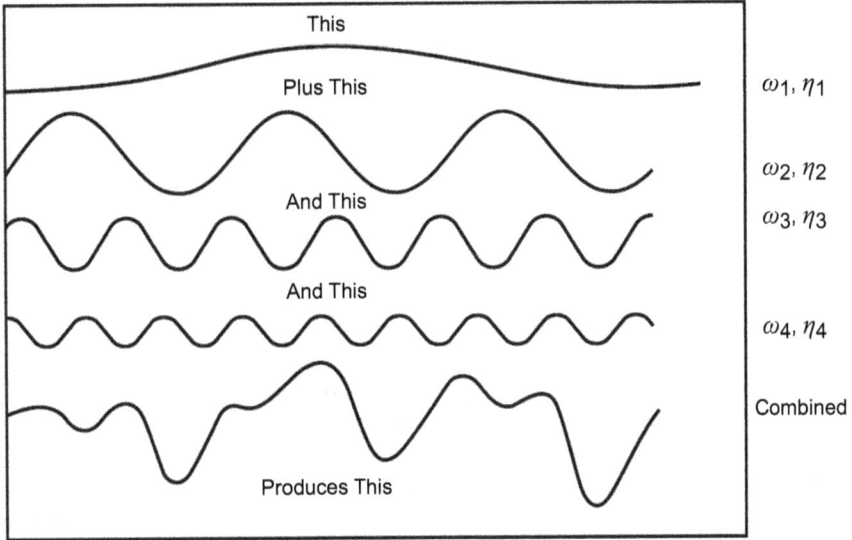

Figure 2.5   Frequency-domain decomposition of irregular pattern combining four regular waves

domain. Phase information can also be obtained in the bottom of Figure 2.5 where positive, zero and negative phase shifts exist in the three decomposed signals.

2.2.4   Wave energy calculation

Wave power has greater power density than wind power. The calculation of the available power from a wave plays an important role in designing a wave energy converter (WEC). Other than that, the knowledge of power possessed by ocean waves provides an assessment on the performance of a converter by comparing the power extracted with the power existing in the waves.

Wave power is calculated by supposing the water depth is much larger than the wave length and is shown below:

$$P = 64\rho g p^2 H_m^2 T \qquad\qquad (2.2)$$

where:

P is wave energy flux per unit of wave front or wave-crest length, with the units W/m.
ρ is water density.
g is gravity acceleration.
H_m is the mean wave height.
T is one wave period.

Thus, wave energy flux is proportional to wave period and the square of mean wave height.

Another measurement of wave energy is wave energy density per unit horizontal area, E, with unit J/m^2, as expressed in (2.3).

The wave's power density, P, is calculated using the wave energy density divided by wave period, which is given in (2.4).

$$E = 1/16\rho g H_s^2 \tag{2.3}$$

$$P = 1/16\rho g H_s^2/T \tag{2.4}$$

where H_s is the specific wave height.

2.2.5   Wave energy spectrum

Ocean waves are produced by the wind passing over the surface of the sea and the amplitude, frequency and the energy of waves depend upon that wind. Larger waves are created by stronger winds or greater exposure to the wind. A wave energy spectrum, which can indicate energy distribution with a range of frequencies at a given wind speed, is an effective method for analysing ocean waves at a specific location (Dean *et al.* 2013).

2.2.5.1   PM spectrum

Pierson *et al.* (1964) first proposed the Pierson-Moskowitz (PM) spectrum by assuming that wind speed is constant for a long time over a long sea surface fetch. This type of sea is known as fully developed. To obtain data, an anemometer mounted on weather ships is used to provide wind speed and take the wave observations. Then, a relationship between wave energy density and wind speed is obtained as shown in (2.5).

$$S(\omega) = \alpha g^2/\omega^5 exp(-\beta(\omega_0/\omega)^4) \tag{2.5}$$

Wave energy density is represented using $S(\omega)$ with the units m^2/Hz. In the PM spectrum, angular frequency $\omega = 2\pi f$, where f is the wave frequency in Hz, $\alpha = 8.1 \times 10^{-3}$, $\beta = 0.74$, $\omega_0 = g/U_{19.5}$ and $U_{19.5}$ denotes the wind speed measured by an anemometer on the weather ship at the height of 19.5 m above the water level.

Equation (2.5) does not show at what frequency the waves have the most energy but it can be simplified by choosing the peak frequency. The peak frequency is obtained by making $dS/d\omega = 0$, which is shown in (2.6), and the PM spectrum can be rewritten in (2.7) by applying the peak frequency ω_{peak} instead of ω_0.

$$\omega_{peak} = 0.877\omega_0 = 0.877g/U19.5 \tag{2.6}$$

$$S = \alpha g^2/\omega^5 exp(-5/4(\omega_{peak}/\omega)^4) \tag{2.7}$$

2.2.5.2   JONSWAP spectrum

In 1973, for the Joint North Sea Wave Project (JONSWAP), Krogstad *et al.* (1999) proposed a new wave energy spectrum by analysing data collected from the island of Sylt, in the Heligoland Bight of North Germany, in the North Sea. It was found that the ocean could not become fully developed because of nonlinear energy transfer between waves and wave interactions, and the interactions between very short and longer waves. To get the JONSWAP spectrum, γ_r is introduced into the PM spectrum as shown in (2.8) and (2.9). It is used to represent wind-wave growth state, with the value range between 1.5 and 6, and it is normally called the peak enhancement factor FINO (2014).

$$S(\omega) = \alpha g^2 / \omega^5 exp(-5/4(\omega_{peak}/\omega)^4) \; \gamma_r \tag{2.8}$$

$$\gamma_r = exp(-((\omega - \omega_{peak})^2/2\sigma\omega_{peak}^2)) \tag{2.9}$$

where:

$\gamma_r = 3.3$
$S = 0.07$ when $\omega \leq \omega_{peak}$
$S = 0.09$ when $\omega \geq \omega_{peak}$

In the JONSWAP spectrum, the constant α and peak angular frequency ω_{peak} are determined as given in (2.10) and (2.11) by measuring u_{10}, which is the wind speed 10 metres above sea level. The fetch, F, is the distance over which the wind blows with a constant velocity:

$$\alpha = 0.076 \left(U_{10}^2 / Fg \right)^{0.22} \tag{2.10}$$

$$\omega_{peak} = 22 \left(g^2 / U_{10}F \right)^{1/3} \tag{2.11}$$

2.2.6   Wave prediction and measurement

Ocean wave forecasting has been widely applied to naval ship design processes and naval operations for a long time. With the help of radar techniques, real-time observations of ocean waves can be achieved (FINO, 2014). For the past few years, 48–96 hour accurate wave forecasting has been provided by the Naval Research Laboratory (2009). Although it is an effective way of forecasting the general ocean waves' situation, it cannot accurately predict detailed information wave by wave. With the rapid development of wave energy, detailed wave-by-wave prediction, such as wave frequency, wave height and wave speed, several seconds into the future are increasingly important for WEC control in real seas.

Since the late 1980s, a number of contributions to future wave prediction have been made. The autoregressive (AR) model is regarded as an effective way to attempt to predict waves, based on the previous outputs. Forsberg (1986) first developed the AR model by adding a moving average to the AR formulation (autoregressive moving average model [ARMA]) to describe ocean waves off the coast of Sweden. Recently, Fusco *et al.* (2009, 2010) proposed linear autoregressive models to predict up to 20 seconds of future waves with good accuracy. It has proved to be a promising approach for predicting future wave elevations from only its past history. However, the application of AR models results in increased instrumentation costs because several distant observations are required to reconstruct a new wave field.

2.3   Potential tidal resource

Tides are a result of the relative motion of the earth, moon and sun, which means they are regular and predictable, with height variations according to season and location. This is one of the attractions of the tidal resource: the timings and the seasonal pattern are known in advance, even if the strength on a particular day cannot be precisely predicted.

2.3.1   Estuarial sources

One strategy is to build barrages across estuaries with a large tidal range. In France, the barrage across the mouth of the River Rance is an experimental 240 MW plant completed in the 1960s. A scheme rated at 254 MW has recently (2011) been completed in South Korea and is said to be the world's largest tidal power plant. The project exploits a 7.9-mile dam, which has existed at the site since 1994, when the government attempted to create a freshwater lake. When Shihwa Lake became polluted, the tidal power plant was proposed as a way of cleaning the water.

Instead of a barrage, it is also possible to place marine current turbines in a tidal stream. A pioneering scheme is that at Strangford Lough where Marine Current Turbines, now owned by Siemens, installed a 1.2 MW, twin-rotor machine. Strangford Lough, however, is exceptional by virtue of its geography, shown in Figure 2.6, which results in very high flows at the Lough mouth. The power available for estuarial tidal streams is considered in the next section.

Various other tidal projects based around estuaries have been proposed; the Severn barrage across that river in the United Kingdom has been the subject of many feasibility studies. Cost, however, has been an issue, as well as environmental concerns.

Harnessing tidal streams is seen to have less impact than building tidal barrages, but the practicalities of implementation at reasonable cost remain a challenge. There are, however, small experimental tidal stream devices in the Bay of Fundy, Canada and in Murmansk, Russia.

If environmental concerns can be overcome, the United Kingdom, France, Canada and Korea would be well placed for tidal barrage schemes, with up to 100 TWh per year potentially available.

2.3.2   Non-estuarial tidal streams

Harnessing tidal streams between islands or off headlands is another possibility, and two approaches can be considered: a tidal fence, which is essentially a barrage; and tidal turbines, built off the seabed or moored. The world resource in terms of tidal streams is shown in Figure 2.7.

In the British Isles, there are attractive tidal sites around Orkney, Shetland, the Severn Estuary, Morecambe Bay and the Channel Islands, see Table 2.4.

Further resources are available worldwide from tidal stream flows of which the United Kingdom could exploit 40 TWh per annum, compared to the UK's current annual electricity consumption of 350 TWh.

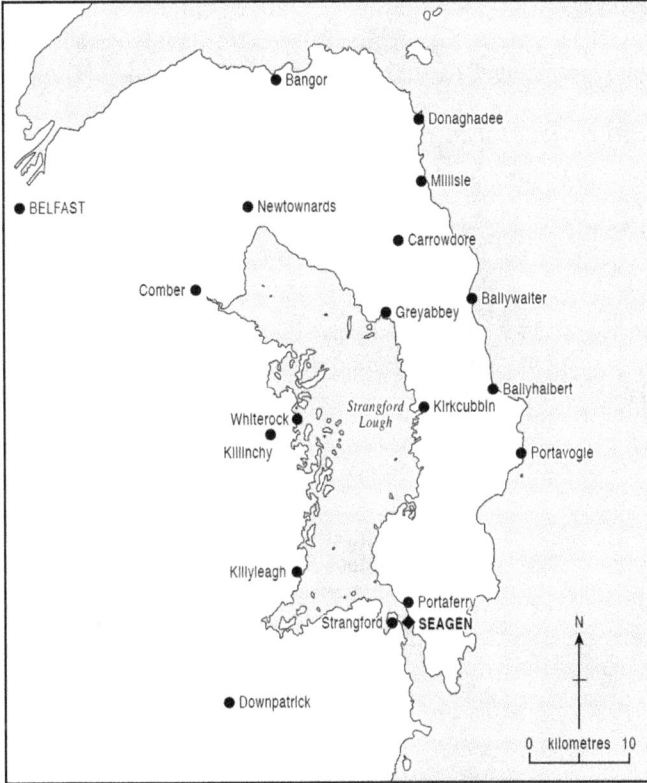

Figure 2.6   Strangford Lough, Northern Ireland

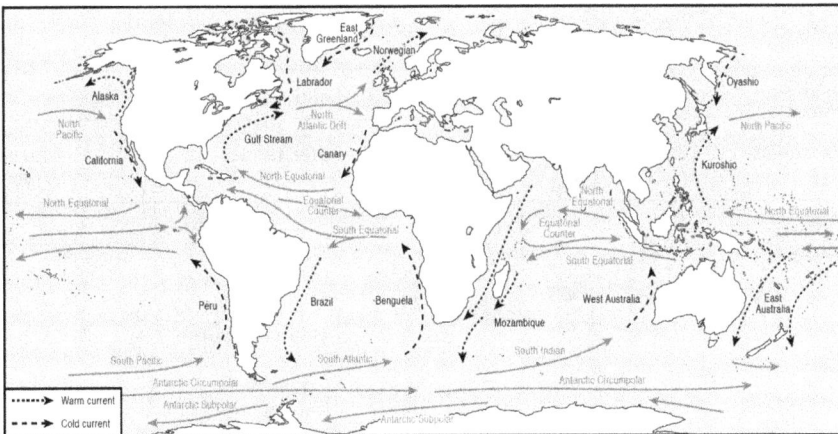

Figure 2.7   World tidal current. Source: marine bio.org

Other countries with good sites include Indonesia, the Philippines, China and Japan. The available resource by region in Western Europe is shown in Table 2.3 and sites around the United Kingdom are listed in Table 2.4.

2.3.3   Tidal characteristics

The power available from a tidal stream is given as follows:

$$P = 0.5\rho A u^3 \tag{2.12}$$

where:

ρ is the density of sea-water, 1030 kg/m^3.
A is the cross-sectional area of the tidal device.
u is the tidal stream velocity.

Table 2.3   Tidal resource by Western European region

Country	Technically available tidal resource		Percentage of European tidal resource
	Power (GW)	Energy (TWh/yr)	
UK	25.2	50.2	47.4
France	22.8	44.4	42.1
Ireland	4.3	8.0	7.6
Netherlands	1.0	1.8	1.8
Germany	0.4	0.8	0.7
Other W European	0	0	0
Total W European	63.8	105.4	100.0

Table 2.4   Tidal resource at various sites around the United Kingdom

Tidal range sites		Tidal stream sites		
Site names	Resource (TWh/yr)	Area	Site names	Resource (TWh/yr)
Severn	17.0	Pentland	Pentland Skerries	3.9
Mersey	1.4	Firth	Stroma	2.8
Duddon	0.212		Duncanshead	2.0
Wyre	0.131		South Ronaldsay 1	1.5
Conwy	0.06		South Ronaldsay 2	1.1
			Hoy	1.4
		Alderney	Casquets	1.7
			Race of Alderney	1.4
		North	Rathlin Island	0.9
		Channel	Mull of Galloway	0.8

Table 2.5   Comparison of the energy available from tidal stream and wind

Energy resource		Marine currents			Wind
Velocity	(m/s)	1.0	2.0	3.0	13.0
	(knots)	1.9	3.9	5.8	25.3
Power density	(kW/m^2)	0.52	4.12	13.91	1.37
Turbine diameter	(m)	15			60

The power density is considerably greater than that for a WT as, although the velocity is low, the sea-water medium is dense. The available power density is perhaps four times that of a WT, so water turbines can be smaller. The data in Table 2.5 illustrates the point: water and WTs are 15 m and 60 m in diameter, respectively.

Favoured tidal sites will have a stream velocity of 2–2.5 m/s. Higher velocities will stress equipment excessively and lower velocities are uneconomic. Sites are ideally within 1 km of the shore to make connection easy, and a water depth of 20–30 m is good. The extractable energy can be up to 10 MW/km^2, implying large arrays for high outputs.

2.4   Resource, location and reliability

Wind, wave and tidal device reliability is dependent upon resource influences, for the following reasons:

(i)   The aerodynamic or hydrodynamic resource is a predictable but complex fluid continuum. Wind resource shows meteorological-scale or synoptic-scale dynamics and fast-fluctuation small-scale turbulence (Milan *et al.* 2014), affecting device forces, torques and power flows. This is exemplified for wind by a conventional image of that process (van der Hoeven 1957), showing a meso-scale, 2D inverse vortex cascade and a short time-scale, 3D, downward vortex cascade, with hypothetical spectral separation (Figure 2.8).

Recent wind studies, however, now propose a universal vortex cascade, containing intermittent fluctuations down to small time scales; a typical multi-fractal process (Fitton 2013).

Tidal resource is also predictable for a given site, but affected by the interaction of wind and wave, each having a probabilistic distribution, and all three depending upon the specific site. Reports from the tidal industry, Delorm (2014), Okorie (2011) and Wood *et al.* (2010), show an increasing understanding of the importance of small timescale turbulent aspects, as in the wind industry. Figure 2.9 demonstrates resource turbulence effects on a TSD, off-shore WT and WEC and their effects on the drive train.

(ii)   A device delivering power, subject to these turbulent resources, experiences a series of irreversible deteriorating processes or failure

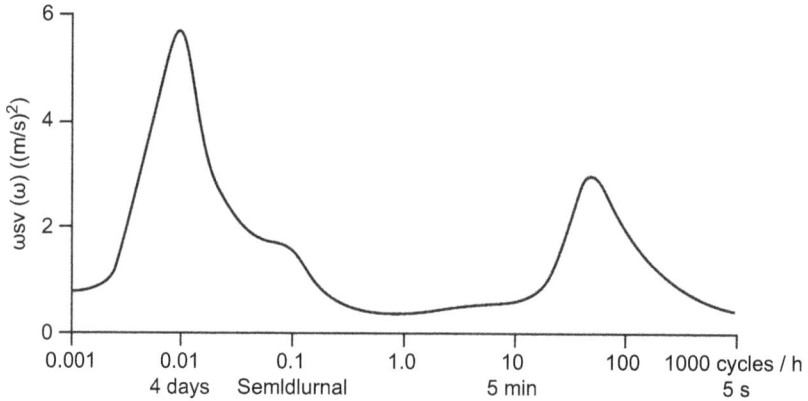

Figure 2.8   Power spectrum of horizontal wind speeds, van der Hoeven (1957)

mechanisms due to rapidly changing forces, torques and power flows. These create stress, thermal and mechanical wear, yield, fatigue and corrosion, which may be controlled by device design but interfere with the energy conversion process.

It is clear from Figure 2.9 that the weakest turbulent effect is probably on the off-shore WT; it is more substantial on a TSD but that a WEC probably sees the largest turbulent excursions;

(iii)   In time, the failure mechanisms of (i) and (ii) accumulate to cause component failures, a stochastic integration process, worsening device reliability and ultimately causing device failure.

Sea measurements, which could illuminate these effects, were made by the FINO project (2014) at the German prototype off-shore wind site, Alpha Ventus, in the Heligoland Bight, near the island of Sylt, in the North Sea, referred to by Krogstad *et al.* (1999). Whilst this was not necessarily a valid wave or tidal MEC site, it enables a closer view of the interaction of wind and wave conditions, Figure 2.10, specific to that site.

An interesting observation is that the wind resource variation distribution is much wider than that of the wave resource, as one would intuitively expect of a less dense and less constrained medium but the denser liquid medium imposes larger turbulent excursions (Figure 2.9).

Seasonal variations in averaged wind speed, wave height and period probabilities will be of interest to MEC farm operators. The normalised, relative variations at the Alpha Ventus site of these three parameters, averaged over an eight year period, are shown in Figure 2.11. These indicate typical North Sea seasonal variations, where 1.0 is the norm.

The impact of the effects of (i), (ii) and (iii) above on the reliability of MECs will be the subject of the following chapter.

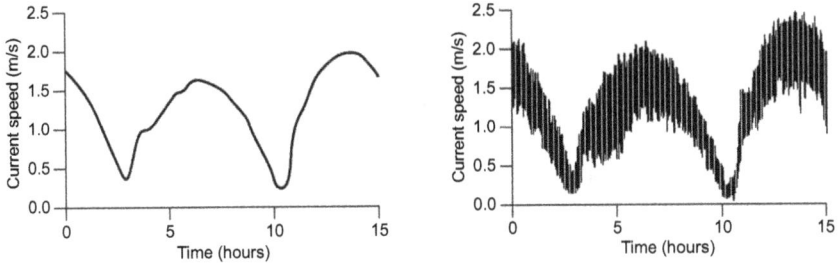

(a)

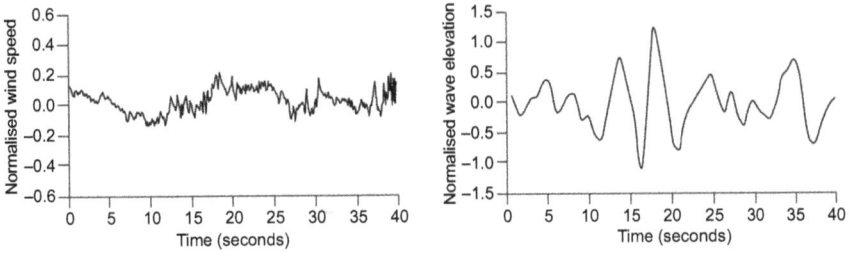

(b)

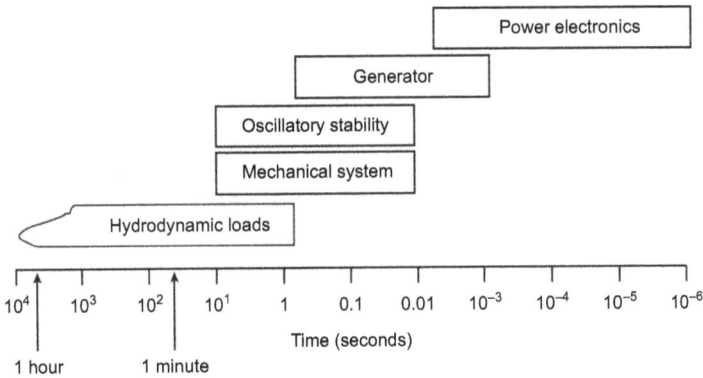

(c)

Figure 2.9 *Effects of turbulence on WTs and MECs: (a) tidal current at a MEC site, smoothed left, and unsmoothed with turbulence right, (b) comparison of effect of resource on an off-shore WT, left, and WEC, right and (c) contribution of turbulent effects to a MEC drive train, cf. Figure 2.8 (Alcorn 2014)*

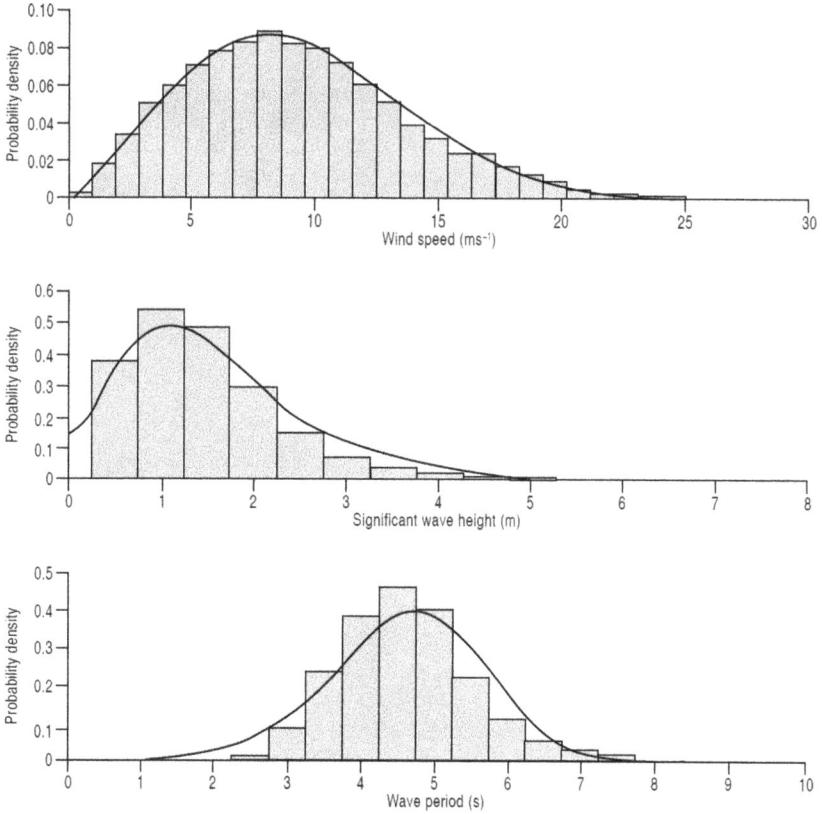

Figure 2.10   Probability distribution of wind speed, wave height and wave period over eight years close to Alpha Ventus site, FINO (2014)

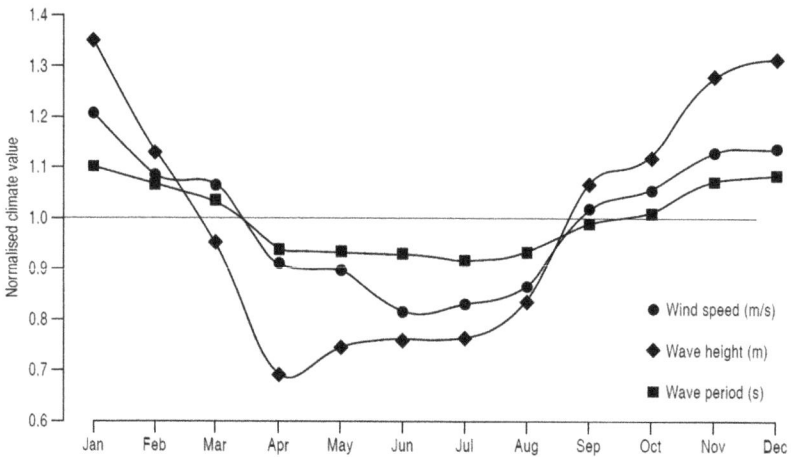

Figure 2.11   Seasonal variation of wind speed, wave height and wave period close to Alpha Ventus site, FINO (2014)

2.5   Summary

Many different ways have been proposed for extracting energy from the sea.

They can be divided into categories by intended location, energy absorption and conversion mechanism.

Prototypes of many tidal and wave systems have been constructed and some have been in operation for some time, but in general for less than 1–3 years, producing interesting data that could inform future development; however, such data is as yet not in the public domain.

While this approach is clarifying the best approaches for tidal stream power, there is still more research needed to understand how to progress reliable wave power.

It is certain, however, that specific wave or tidal systems will be selected according to the MEC farm sites available and their specific resource levels and local turbulence will affect reliability.

Design reliability harmonisation will not occur until significant GW capacity has been operated in the sea.

Wave and tidal resource prediction is a complex time- and frequency-domain issue which has a huge impact on the reliability and consequent success of marine devices. This will be addressed in the next chapter.

Chapter 3

Reliability theory relevant to wave and tidal devices

3.1 Introduction

Previous chapters described the background to wave and tidal energy resource and MEC extraction; we now need to understand the principles of reliability that can be applied to these devices.

Reliability analysis must be informed by causality and the author's view is that one of the important lessons learnt in energy extraction over the last 40 years has been to consider reliability firmly in terms of measured machine failure rates and their root causes. In so doing, successful reliability analysis will address machine unreliability directly or, more importantly for the operator, the device availability.

This chapter sets out the reliability theory relevant for MECs; however, the amount of reliability data available for such devices is severely limited. The theory below is set out with this in mind so that MEC innovators, developers and operators can collect the appropriate data, see how reliability can influence their machines and based on the data make appropriate design, deployment and maintenance decisions.

A 1–2 MW MEC is a complex, robotic, electro-mechanical generating machine mounted in or on a steel/concrete structure, with a steel or concrete foundation or moored to steel anchors or concrete foundations.

From such a device, reliability is dependent on epistemic uncertainty, Sorensen (2014), affecting:

1. Structural reliability for which predicted failure rates are probably $<10^{-4}$ failures/yr and the probabilistic spread of those low failure events needs to be considered, characterised by limit state equations defining failure or unacceptable behaviour.
   (a) These are based upon an MTBF much greater than the life of the device, where design must be based upon predicting the extreme events the structure must withstand;
   (b) Failure to predict these accurately can lead to over-design and the risk of increasing device CAPEX.
2. Electro-mechanical reliability, ranging from 1 to 10^{-3} failures/yr, subject to the normal vagaries of machinery and can be predicted using classical

reliability methods such as estimated constant failure rates for individual sub-assemblies.

(c) These are based, as we shall see, upon probabilistic distributions about a mean MTBF less than the life of the device, where components are essentially repairable.

(d) The problem here is that the operator may not be efficient at replacing devices when repair is needed or operational access is difficult and this increases device OPEX.

- Control system reliability, which depends upon the environment, electro-mechanical issues and the reliability of the software contained within the control system. However, the lesson from other power generation and aerospace devices is that this is an area where startling improvements in reliability can and will be progressively achieved.

Reliability analysis of MEC systems is more complex than for other generating machinery, but similar to WTs, because they also are subject to aleatory uncertainty, see Section 2.4, due to:

- Waves and tides from which the device extracts energy;
- Stochastic effects of wind, waves and weather on the device;
- Effects of device corrosion.

In order to understand and predict these effects, there must be a detailed understanding of reliability theory, a relevant textbook on the subject is Birolini (2007).

3.2   Background

3.2.1   Terminology

A full terminology is given at the start of this book.

The reliability of a sub-assembly in a MEC is defined as the probability that it will meet its required function under stated conditions for a specified period of time. This definition of reliability has four essential elements:

- Probability;
- Required function;
- Time variable;
- Operational conditions for adequate performance.

Unreliability is the complement of reliability and is related to a failure intensity function, $\lambda(t)$, to be defined later. Failure is the inability of a part to perform its required function under defined conditions. The part is then in a failed state, in contrast to an operational or working state.

Reliability is also sometimes expressed as the mean time between failures, *MTBF* or θ and it can be shown that $\theta = 1/\lambda$, where λ is the special condition when

$\lambda(t)$ = constant and the failure intensity can be expressed as a failure rate. Failure rates are useful because they can be compared between different sub-assemblies and so the reader can ascertain which sub-assemblies will have the most effect on overall device reliability.

This reliability definition experiences difficulties as a measure for continuously operated systems such as MECs that tolerate failures and can be repaired.

Then a more appropriate measure is availability, A, defined as the probability of finding the system in the operating state at some time into the future. This definition then reduces to only two elements:

- Operability;
- Time variable.

A non-repairable system is one which is discarded after failure; examples are small batteries or light bulbs.

A repairable system is one that, when a failure occurs, can be restored into operational condition after any action of repair, other than replacement of the entire system; examples are WTs, MECs, car engines, electrical generators and computers.

Repair actions can be an addition of a new part, exchange of parts, removal of a damaged part, changes or adjustment to settings, software update, lubrication or cleaning. The following sections will deal with reliability, MTBF, $\theta = 1/\lambda$.

Issues of availability will be dealt with later in the book in Chapter 6.

3.2.2 Failure mechanisms

The process from operation to failure for a specific failure mode in a typical sub-assembly or component contains the root cause-failure mode-failure sequence essential to understanding reliability. An example is given in Figure 3.1 for the main drive shaft of a MEC device.

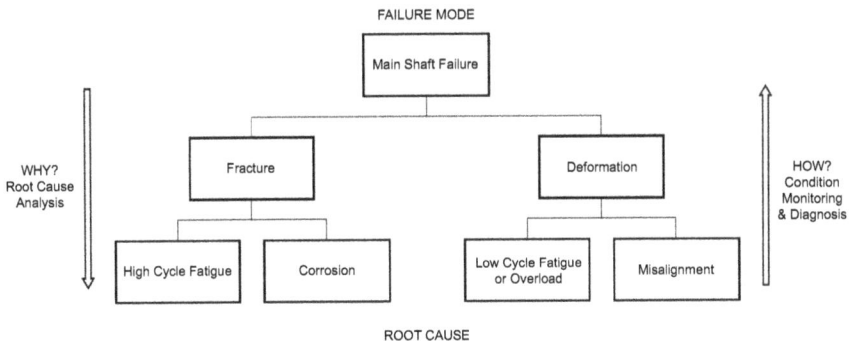

Figure 3.1 Relationship between failure sequence and root cause analysis (RCA) for a particular failure, main shaft fracture, also called a cause and effect diagram

The duration of the failure sequence depends on the failure mode, the operating condition of the device and the ambient condition in which it is operating.

Figure 3.1 demonstrates the process of a failure mode for the fracture of a component, a main shaft.

Figure 3.2 plots the normally distributed failure probability density function of failure of another component, a generator. This has a mean set as the generator MTBF equal to approximately the half-life of the system. The consequent survivor function of the generator is also shown.

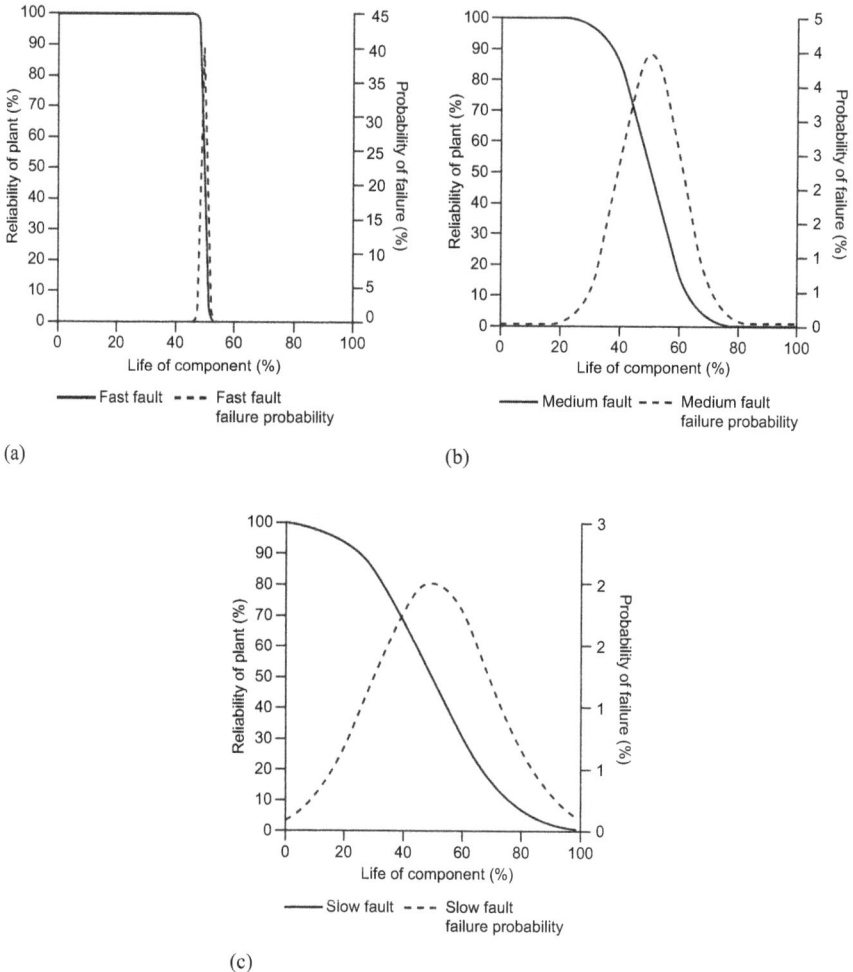

(a)

(b)

(c)

Figure 3.2   Continuous failure probability density function and survivor function for a generator, showing operability falling with time as different faults progress. (a) Fast speed fault, (b) medium speed fault, and (c) slow speed fault.

Figure 3.2(a) shows the progression from reliable to non-reliable operation for a rapid fault occurring rapidly at the 50% life point. The probability of failure rises sharply close to this point, the area under the curve being equal to 1 because in the life of the device there is 100% probability of failure. This might represent a generator winding electrical fault.

Figure 3.2(b) repeats the sequence for a medium fault and this might represent, for example, the probability of a generator rolling element bearing fault.

Finally, Figure 3.2(c) repeats the sequence for a slow fault, which might represent, for example, the slow deterioration in the generator ventilation system.

Understanding these processes lies at the heart of understanding device reliability.

If a failure sequence is very rapid, for example a few seconds, like the fast fault in Figure 3.2(a), then effective detection and remedial action are impossible. This is the situation for most electrical failure modes which actuate electrical protection, where the period of action of the final failure mode may be only a few seconds or even only a few cycles of the generator frequency.

However, if the failure sequence is days, weeks or months, like the slow fault in Figure 3.2(c), then monitoring has the potential to provide early warning of impending failure and the ability to continue operating the device before failure and then maintain it to avoid failure. Therefore, improving device reliability must concentrate on:

- Removing Root Causes and Failure Modes that exhibit rapid fault sequences, Figure 3.2(a), for example by designing or using reliable, fast-acting electrical fault protection;
- Devising a means for detecting Root Causes and Failure Modes that exhibit Failure Sequences of substantial duration, such as Figure 3.2(b) and (c), and carrying out maintenance or replacement action to mitigate their impact.

A valuable technique for establishing the failure modes, root causes and failure mechanisms of machinery is failure modes' effects and criticality analysis (FMECA). As yet no FMECA has been published for a MEC, although AME (1992) and YARD (1980) are close to it but rather dated, however a useful FMEA guide prepared for WTs by Arabian-Hoseynabadi *et al.* (2010) and this subject will be discussed later in the book in Section 9.1.2.3.

3.2.3   *Reliability block diagrams and taxonomy*

Individual MEC sub-assemblies exhibiting faults can be represented by reliability block diagrams (RBD) in a set and then connected in series or parallel to represent their functionality. Figure 3.3 shows possible arrangements for two reliability blocks.

3.2.3.1   **Series systems**

From a reliability point of view, parts in a set are said to be in series if they must all work for system success and only one needs to fail for system failure. Consider a

(a)

(b)

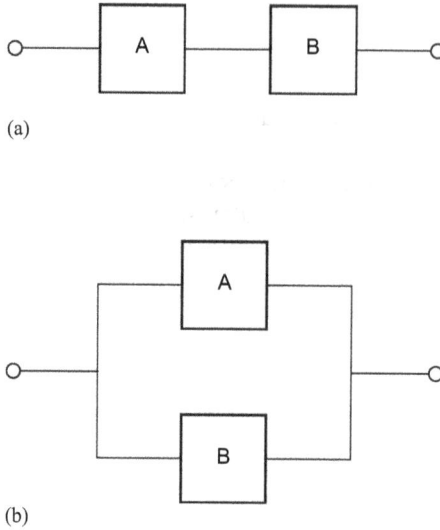

Figure 3.3   Representation of sub-assemblies in a reliability block diagram.
(a) Series components and (b) parallel components.

system consisting of two independent parts A and B connected in series, for example a gear train.

Let R_a and R_b be the probability of successful operation of the individual parts A and B, respectively, in Figure 3.3(a) and R_s be the probability of successful operation of the series set.

$$R_s = R_a * R_b \tag{3.1}$$

Or more generally:

$$R_s = \Pi \; R_i$$

This is the reliability product rule.

Let Q_a and Q_b be the probability of failure of sub-assemblies A and B, respectively:

$$Q_s = 1 - \Pi \; R_i \tag{3.2}$$

Example:

A gearbox consists of six successive identical gear wheels, all of which must work for system success. What is the system reliability of the series set if each gearwheel has a reliability of 0.95? From the product rule:

$$R_s = 0.95^6 = 0.7350$$
$$Q_s = 0.2650$$

3.2.3.2   Parallel systems

Parts in a set are said to be in parallel, from reliability point of view, if only one needs to be working for system success or all must fail for system failure.

Consider a system consisting of two independent parts A and B, connected in parallel, Figure 3.3(b), for example two lubrication oil pumps connected in parallel for a gearbox. From a reliability point of view, the requirement is that only one sub-assembly has to be working for system success.

Again let R_a and R_b be the probability of successful operation of individual sub-assemblies and R_p be the probability of successful operation of the parallel set. Let Q_a and Q_b be the probability of failure of sub-assemblies A and B, respectively:

$$R_p = 1 - Q_a * Q_b \qquad (3.3)$$

More generally:

$$R_p = 1 - \Pi Q_i \qquad (3.4)$$

$$Q_p = \Pi \ Q_i \qquad (3.5)$$

Example:

A system consists of four pumps in parallel each having reliabilities of 0.99. What is the reliability and unreliability of the parallel set?

$$Q_p = (1 - 0.99)^4 = 0.01^4 = 0.00000001$$

$$R_p = 1 - Q_p = 0.99999999$$

3.2.3.3   Device configurations and MEC taxonomy

The previous sections show that device configuration is an important factor in the cause and effect diagram and in the aggregation of failure modes.

We may consider that configuration to be simply the assembly configuration of individual components into sub-assemblies and then the aggregation of those sub-assemblies into a complete device.

Billinton *et al.*(1992) made clear that the relevant configuration is not necessarily the assembly diagram but that due to the reliability dependence of components, where the failure of one component in the configuration may not incur the failure of the device, for example due to redundancy. This affects the makeup of the cause and effect diagram, which is the configuration relevant to condition monitoring, and is governed by causality rather than assembly.

A sub-assembly reliability configuration can be built up from the cause and effect diagrams of individual components to give a cause and effect diagram for the sub-assembly such as shown for a main shaft in Figure 3.4.

On the other hand, an assembly configuration is shown in Figure 3.5 for a typical MEC induction generator. An assembly configuration may be adequate for reliability consideration in some simplified failure cases but is not necessarily correct for the assessment of failure modes. To distinguish between the two approaches, a designer must do an FMECA.

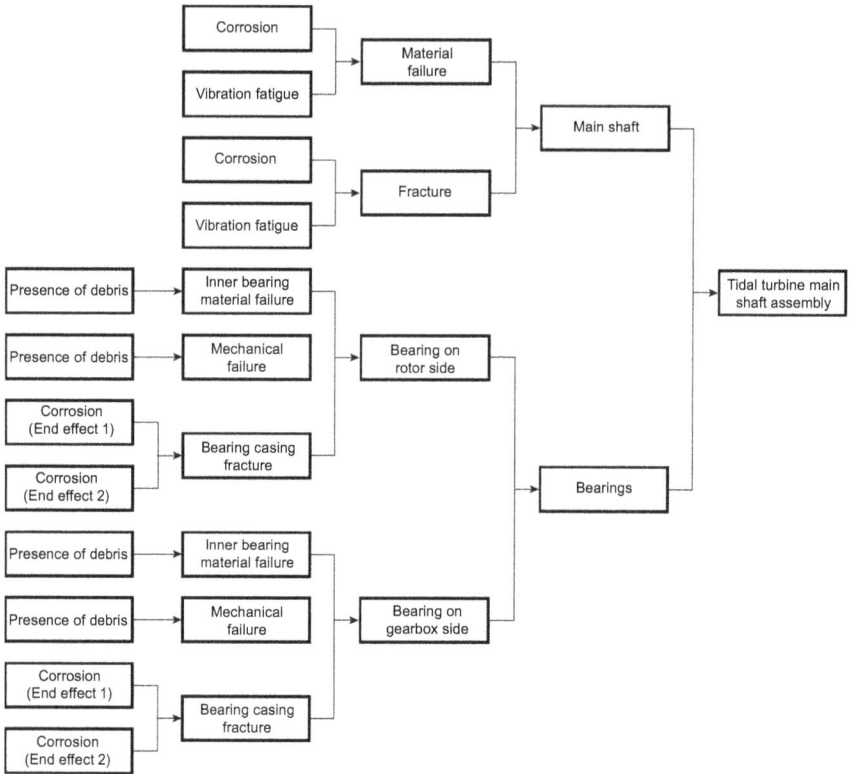

Figure 3.4   Example of a reliability cause and effect diagram for a typical MEC sub-assembly, the main shaft

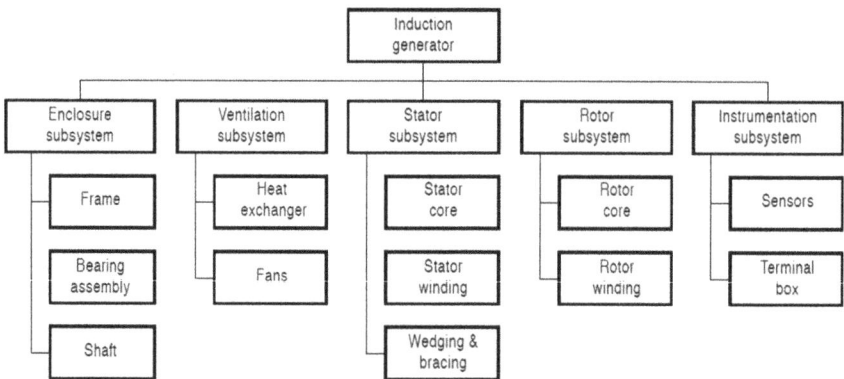

Figure 3.5   Example of an assembly configuration for a typical MEC sub-assembly, an induction generator

A simple example of this issue could be the lubrication pumps of a large turbo-generator. It is customary to introduce redundancy into this plant by providing:

- 2 off, 100%-rated, main and standby, AC motor-driven lubrication pumps;
- With a third 100%-rated, DC motor-driven pump provided with emergency power from a battery, when AC power fails.

In this case, it is not legitimate simply to add the cause and effect diagrams in series or to add the failure mode probability functions of the three pumps, since a large turbo-generator can operate with any one out of three pumps functioning. The correct approach is to consider the three systems in parallel and take account of the probability density functions in the same way, see Section 3.2.3.

Behind the simple series and parallel block examples above lies the potential to increase device redundancy and secure higher reliability. A real MEC is made up of a complex combination of series and parallel block sets, interacted upon by the device controller. However, most generation devices are dominated by series block sets, limiting the potential for redundancy.

The development of WECs & TSDs has thrown up a number of innovative taxonomies, for example the Pelamis WEC and the Seagen TSD, to be introduced later, which incorporate degrees of parallelism, which deliver higher intrinsic reliability, as will be shown later.

Therefore, this structure or MEC taxonomy is crucial to their potential reliability and therefore to their availability and ultimately the CoE they deliver and this will be the subject of Chapter 6.

3.2.4   Root causes

Section 2.4 referred to the failure mechanisms inside a MEC device, some of which are particular to the marine resource. Before considering the details of reliability calculation, it is useful to try and itemise these failure modes as follows and then consider whether they could be predicted mathematically:

- Thermal integrity
- Electrical integrity
- Mechanical wear
- Material integrity
  - Stress
  - Yield
  - Low cycle fatigue
  - High cycle fatigue
- Corrosion

Component failure modes and the failure mechanisms are controlled by statistical processes in the components, such as:

- Insulation degradation in electrical windings;
- Degradation of metal components in bearings or shafts due to fatigue.

These processes govern the transition from operation to failure, the subject of detailed texts on reliability such as Billinton *et al.* (1992). Each component failure mode will have its own probability density function, $f(x)$, that derives from its own physical processes.

For example:

- The shape of the probability density function for winding insulation degradation depends upon ambient and operating temperatures, deteriorating slowly as the insulation degrades with time;
- Whereas the probability density function for bearing or shaft fatigue would show a rapidly rising trend to failure as fatigue cycles are accumulated;
- A particular problem occurs with complex fatigue loadings, partially resolved by Miner's rule, Wikipedia (2016), where there are k different stress magnitudes, S_i, in a spectrum, S_i *($1 \leq i \leq k$)*, each contributing $n_i(S_i)$ cycles, then if $N_i(S_i)$ is the number of cycles to failure of a constant stress reversal S_i, failure will occur when:

$$\sum_{i=1}^{k} (n_i/N_i) = C \tag{3.6}$$

C lies experimentally between 0.7 and 2.2; for design purposes it is often assumed to be 1. This can be thought of as assessing the proportion of life consumed by a linear combination of stress reversals at varying magnitudes.

Though Miner's rule can be a useful approximation in some circumstances, it has several major limitations:

- It fails to recognise the probabilistic nature of fatigue and there is no simple way to relate life predicted by the rule with the characteristics of a probability distribution. Industry analysts use design curves, adjusted to account for scatter, to calculate $N_i(S_i)$;
- There is sometimes an effect due to the order in which the reversals occur:
  - In some circumstances, cycles of low stress followed by high stress cause more damage than would be predicted by the rule;
  - In the event of an overload, this may result in a compressive residual stress, retarding crack growth; therefore cycles of high stress followed by low stress may cause less damage, due to the presence of this residual compressive stress;
- Therefore, it may be unwise to invoke the use of Miner's rule for the types of fatigue encountered in marine turbine and wave devices, see Section 2.4.

The mathematical function, $f(x)$, for a component failure mode density function should have the flexibility to represent a wide range of major failure modes as present in most machinery. Therefore, different distributions can model different failure mechanisms, which are the essential physics of failure linking cause and effect and in MECs including, for example:

- Elimination of early teething problems;
- Hydrodynamic and aerodynamic mechanisms:
  - Fatigue due to aero-elastic and hydro-elastic behaviour;
  - Fatigue failure due to wind and water turbulence;
  - Failure due to sudden wind gusts, tidal flow excursions or wave impacts.

- Mechanical mechanisms:
  - Fatigue due to gear meshing;
  - Fatigue activity in bearings;
  - Mechanical wear in gearing and bearings.

- Electrical mechanisms:
  - Thermal ageing;
  - Thermo-mechanical cycling fatigue in electromechanical components as MECs come on and off load;
  - Thermo-mechanical fatigue stress in power electronic components during normal variable load operation.

- Corrosion mechanisms:
  - Corrosion of structures, anchors and foundations;
  - Corrosion of electrical connectors.

- Operational factors such as lack of the staff or plant to change components known to be close to the limit of their working range;
- The physical time needed to change of parts.

Each can be represented by different probability distributions. The ability to distinguish between them allows faults to be detected and distinguishing these processes requires an understanding of reliability theory.

3.3 Theory

3.3.1 Reliability functions

We now present some simple reliability functions so that we can understand the random and continuous variables important to MEC reliability; in each case we have taken the random variable x to be time t as this simplifies the understanding. Therefore, the functions we consider are $f(t)$, rather than probability functions $p(x)$ where x could be any random plant variable.

The following equations and mathematical relationships between the various reliability functions do not assume any specific failure distribution and are equally applicable to all probability distributions used in reliability evaluation. Consider N_0 identical non-repairable parts are tested:

$$N_s(t) = \text{number surviving at time } t \tag{3.7}$$

$$N_f(t) = \text{number failed at time } t \tag{3.8}$$

$$\text{Therefore } N_s(t) + N_f(t) = N_o \tag{3.9}$$

At any time t, the survivor or reliability function, $R(t)$, is given by:

$$R(t) = N_s(t)/N_o \qquad (3.10)$$

Similarly, the probability of failure or unreliability function, $Q(t)$, is given by:

$$Q(t) = N_f(t)/N_o \qquad (3.11)$$

where:

$$R(t) = 1 - Q(t) \qquad (3.12)$$

See Figure 3.6 and the failure density function $f(t)$ is given by:

$$f(t) = 1/N_0 \left(dN_f(t)/dt\right) \qquad (3.13)$$

The relationship between failure density function $f(t)$ the survivor or reliability function, $R(t)$, and the unreliability function, $Q(t)$, is:

$$f(t) = dQ(t)/dt = -dR(t)/dt \qquad (3.14)$$

or

$$Q(t) = \int 0 \, tf(t)dt \qquad (3.15)$$

and

$$R(t) = 1 - \int 0 \, tf(t)dt \qquad (3.16)$$

The total area under the failure density function must be unity, see Figure 3.6. Therefore:

$$R(t) = \int 0\infty f(t)dt = 1 \qquad (3.17)$$

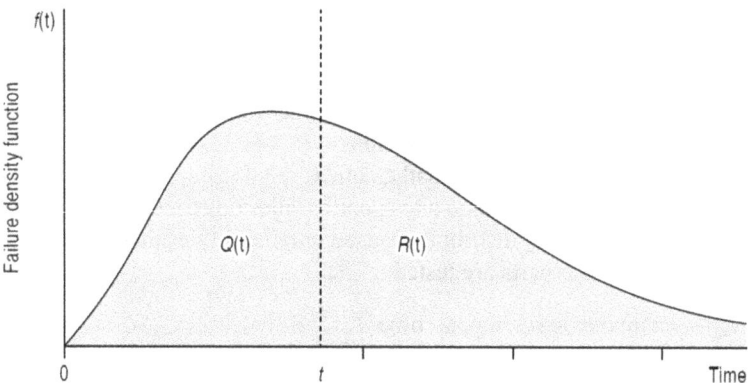

Figure 3.6   Failure density function against time showing reliability R(t) and Q(t)

This also applies to the failure density functions shown in Figure 3.2.

The failure intensity or hazard rate function, $\lambda(t)$, is the failure density function, $f(t)$ Figure 3.6, normalised to the number of survivors:

$$\lambda(t) = (dNf(t)/dt)/Ns(t) \tag{3.18}$$

$$\lambda(t) = (dR(t)/dt)/R(t) \tag{3.19}$$

The normalised hazard rate, $\lambda(t)$, is exceptionally useful for an engineering understanding of how the plant behaves.

The special case in which $\lambda(t)$ is constant and independent of time is an exponential distribution of $f(t)$ and the hazard rate becomes the failure rate and $\lambda(t) = \lambda$ the number of failures per unit time.

3.3.2   Example of reliability data

The following is an example of these methods using contrived data, based upon an example given in Billinton *et al.* (1992), which could be considered as a large renewable farm of 1,000 wave or tidal MECs. For the sake of this example, each is considered non-repairable and there is a steady failure of these devices. Table 3.1

Table 3.1   Record of failures of 1,000 non-repairable MECs in a marine energy extraction farm, based on Billinton et al. (1992)

Time interval, years	Number of failures in each interval, N	Cumulative failures, N_f	Number of survivors, N_s	Failure density function, $f(t)$	Unreliability function or cumulative failure distribution, $Q(t)$	Reliability or survivor function, $R(t)$	Failure intensity or hazard rate, $\lambda(t)$
0	240	0	1000	0.240	0.000	1.000	0.240
1	140	240	760	0.140	0.240	0.760	0.184
2	90	380	620	0.090	0.380	0.620	0.145
3	58	470	530	0.058	0.470	0.530	0.109
4	40	528	472	0.040	0.528	0.472	0.085
5	23	568	432	0.023	0.568	0.432	0.053
6	18	591	409	0.018	0.591	0.409	0.044
7	13	609	391	0.013	0.609	0.391	0.033
8	13	622	378	0.013	0.622	0.378	0.034
9	13	635	365	0.013	0.635	0.365	0.036
10	16	648	352	0.016	0.648	0.352	0.045
11	18	664	336	0.018	0.664	0.336	0.054
12	20	682	318	0.020	0.682	0.318	0.063
13	30	702	298	0.030	0.702	0.298	0.101
14	60	732	268	0.060	0.732	0.268	0.224
15	63	792	208	0.063	0.792	0.208	0.303
16	65	855	145	0.065	0.855	0.145	0.448
17	70	920	80	0.070	0.920	0.080	0.875
18	10	990	10	0.010	0.990	0.010	1.000
19	0	1000	0	0.000			
Totals	1000						

records the cumulative failures and survivors over a period of 20 years calculating the failure density function, $f(t)$ which sums to 1 and the hazard rate, $\lambda(t)$. So Table 3.1 records the reliability of this off-shore MEC farm while Figure 3.7 plots all these functions so that their nature can clearly be seen.

Figure 3.7(c) is interesting as it shows the failure density function, $f(t)$, the area under which accumulates to 1, compared to Figure 3.6.

Figure 3.7(d) is also interesting as it shows the hazard rate, $\lambda(t)$, the failure density function, $f(t)$, normalised to the remaining devices. The hazard rate clearly shows a bath-tub form, which will be discussed later, with the early failures phase I, steady failure rate phase II and wear-out phase III.

Particularly interesting in Figure 3.7(c) is phase II where it shows the failure density function decreasing roughly exponentially, representing the random nature of failures in that phase.

When the failure density function, $f(t)$, is normalised into the hazard rate, $\lambda(t)$, those random failures during phase II become a roughly constant hazard or failure rate, λ.

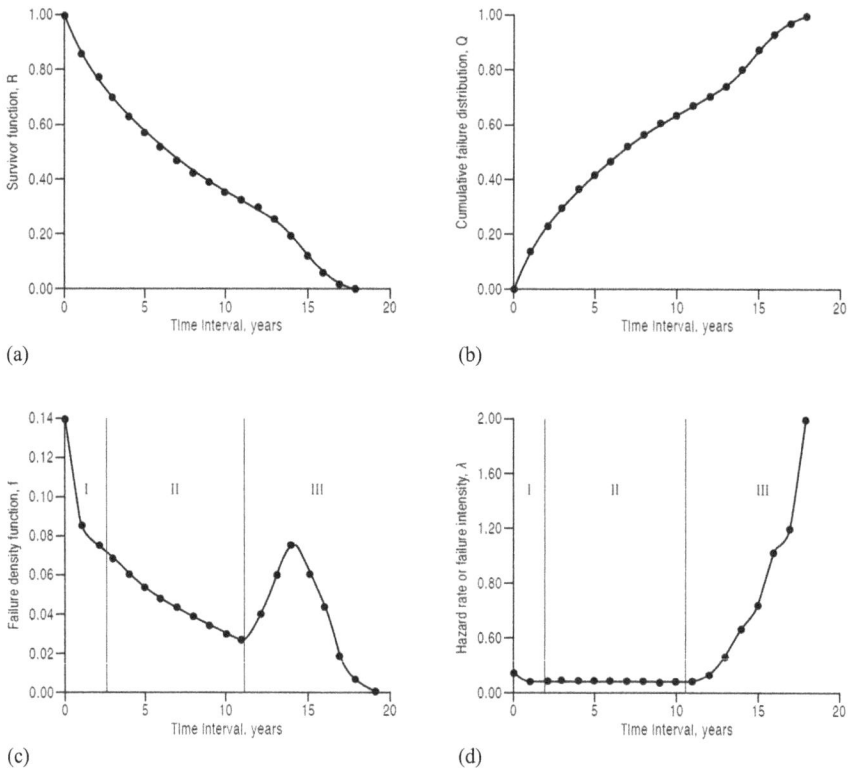

Figure 3.7   Reliability functions from a farm of 1,000 non-repairable devices over 20 years: (a) reliability or survivor function, R(t), (b) unreliability or cumulative failure distribution, Q(t), (c) Failure density function, f(t), (d) Hazard rate or failure intensity, λ(t)

The next section will consider reliability distributions in detail so that we can have an understanding of their application to MECs; however, before that we need to consider the form of variables that can be evaluated to understand reliability.

3.3.3 Random and continuous variables

In order to understand the issues raised in Section 3.2, it is necessary to study the effect of the random and continuous variables associated with failure.

The continuous variable may be recorded continuously or discretely in the context of MEC reliability as the probability of failure p or P, respectively, varying from 0% to 100%. These are recorded against a random variable, x or X, and this variable could be time. Is it always appropriate to use time as the random variable when considering renewable energy devices?

Calendar time may be convenient for recording when a failure occurs but is not necessarily best for reliability analysis, for example the following random variables may be more appropriate;

• Time on test, which excludes time shut down for maintenance of lack of resource, seems more useful;
• Turbine rotations may also be more appropriate, especially for the reliability of aerodynamic or transmission sub-assemblies;
• Energy generated by the MEC, GWh, may also be more appropriate, especially for the reliability of electrical sub-assemblies.

Operators usually cannot measure Time on Test because they cannot easily keep track of the date of origin of the MEC or its parts but they can easily measure the number of failures in an interval of time, for example monthly, quarterly or annually, each failure incident representing 100% probability of failure. This is called censored, discrete data because in that case, time becomes the discrete random variable.

An example of such differences in the plotting of the continuous and random variables is shown in Figure 3.8 where identical WT failure data from the large German WSD survey, referred to in Section 5.3.1.1, are in this case plotted against:

• Discrete censored calendar time, as failure/turbine/year (Figure 3.8(a)).
• Or continuous variable GWh generated, as failure/turbine/GWh (Figure 3.8(b)).

The former shows a reduced failure rate with calendar time. The latter shows a wider variance but an increasing number of failures per GWh generated. This is probably showing the extent of small but significant failures occurring in the growing number of larger, more technically complex WTs.

What this shows us is that the method of collecting and presenting reliability data is important, in particular the choice of the random or discrete variable, x or X, against which continuous variable data collected is important;

• The random variable, x or X, can be presented in different ways but against time is very attractive because time information is easy to collect from many engineering monitoring devices;

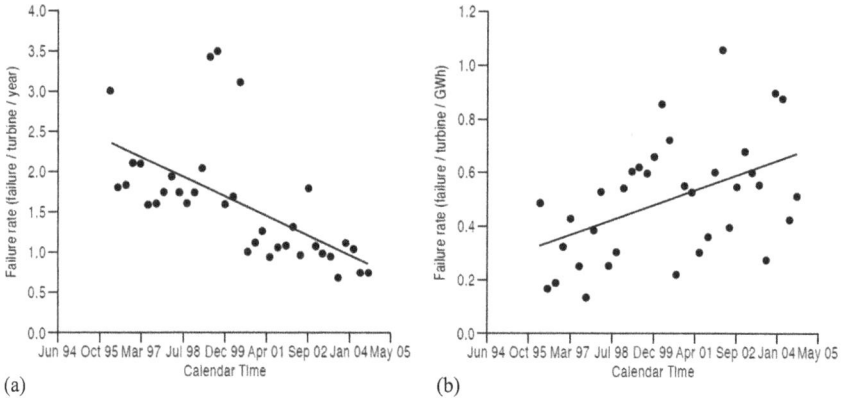

Figure 3.8   Comparison of plots of identical failure data from a Geman WT survey plotted as failures/turbine/year or as failures/turbine/GWh, Tavner (2012): (a) failure/turbine/year vs time, (b) failure/turbine/ GWh vs time

- The random variable can be presented in different ways, whether it is continuous, such as time on test, GWh or rotations, or discrete, such as censored failure data, needs to be decided based on the interpretation needed;
- Plotting the random variable, x or X, in different ways against different continuous variables reveals different information;
- Whether the component, on which the data is being collected, is repairable or non-repairable;
- If the data collection method is good and the continuous variable, p or P, is chosen appropriately, then the statistical data collected should yield robust reliability information. If not the reliability information may be faulty.

The following section will show how some reliability functions are developed and give an example of how they can represent the reliability of off-shore devices.

3.4   Distributions, point processes, Weibull and the bath-tub curve

3.4.1   *Probability distribution function*

Consider an example where there is a probability that a MEC will experience failures during its life cycle. Individual failures are the discrete random variable being counted against X intervals of the life, e.g. months.

The probability of each failure during the first five months of operation could be determined experimentally from the field. Suppose that these probabilities are:

$$P(X = 1) = 0.6561$$
$$P(X = 2) = 0.2916$$
$$P(X = 3) = 0.0486$$
$$P(X = 4) = 0.0036$$
$$P(X = 5) = 0.0001$$

Note that the data is censored into five equal monthly periods, starting from first month.

This gives the probability distribution function (PDF), a graphical representation of a PDF of failures $f(X)$ against X is shown in Figure 3.9. Figure 2.10 also shows examples of PDFs, in that case for the wave height, wave period and wind speed at a particular site.

Definition:

The probability distribution function of the discrete random variable is the probability of failure in each specified censored interval of the variable X. For a discrete random variable f with n possible values at $x_1, x_2, \ldots x_n$, therefore the probability mass function, $f(x_i)$, is expressed as:

$$f(x_i) = P(X = x_i) \tag{3.20}$$

3.4.2 Cumulative distribution function

It is useful to be able to express the cumulative probability such as $P(X \leq x)$ in terms of a formula, the formula for an accumulation of probabilities is called a cumulative distribution function (CDF) and a discrete CDF is shown in Figure 3.10.

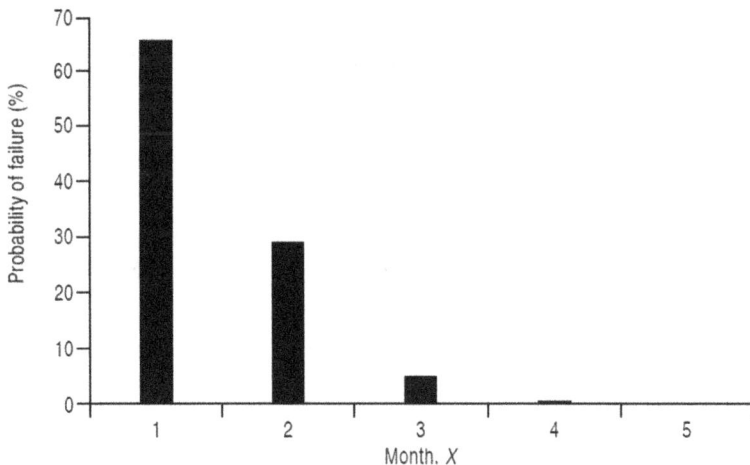

Figure 3.9 Probability distribution function

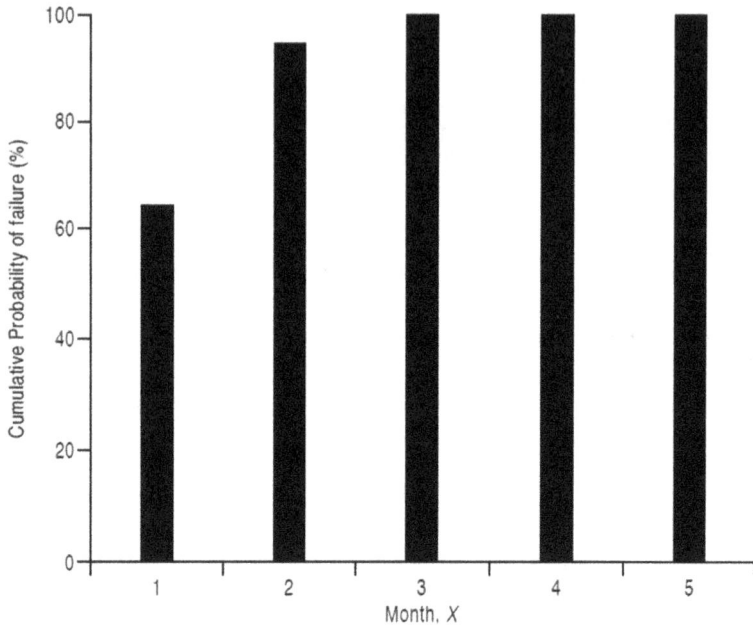

Figure 3.10   Cumulative distribution function

Figure 3.7(b) is an example of a continuous CDF.

Definition:

The cumulative distribution function, $F(x)$, is an analytical method for describing the probability distribution function of a discrete random variable is expressed as:

$$F(x) = P(X \leq x) = \sum f(x_i), \quad \text{for } x_i \leq x \tag{3.21}$$

3.4.3   Discrete binomial function

Consider carrying out a random experiment consisting of n repeated and independent trials:

• Each trial results in only two outcomes, 'success' or 'failure';
• The probability of a success in each trial, p, remains constant.

An example of such a random experiment could be the monitoring of failures, i.e. operation or non-operation at predefined intervals of a critical component such as a bearing in a gearbox, see Figure 3.11.

Definition:

The random variable X that equals the number of trials that result in a success has a binomial distribution with parameters p and $n = 1, 2, 3, \ldots$ The probability

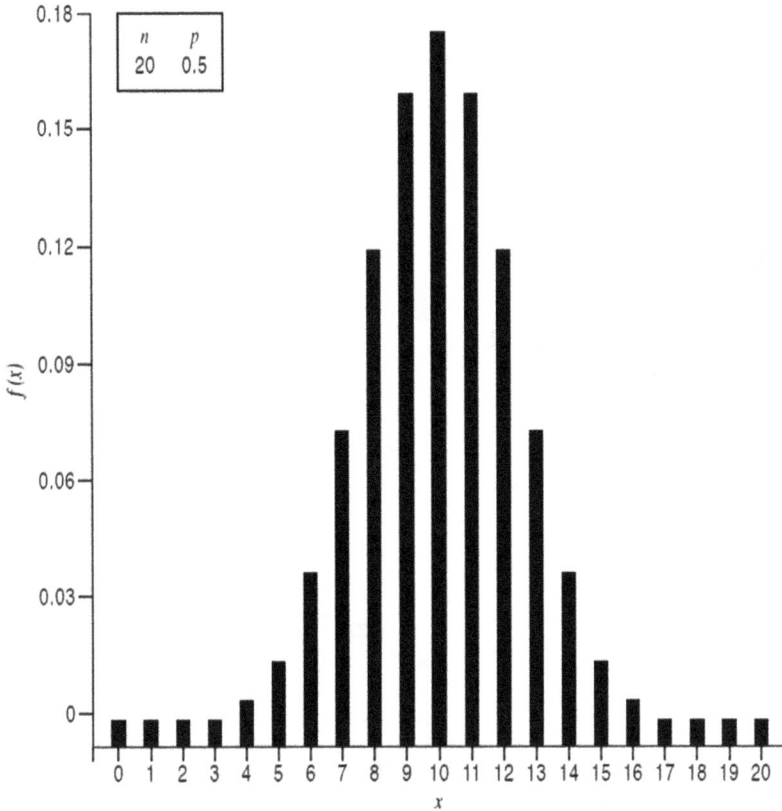

Figure 3.11   Binomial distribution for selected values of n and p

mass function of X is:

$$F(X) = (nx)p^x(1 - p)^{n-x}, \text{ for } x = 0, \ 1, .., n \qquad (3.22)$$

3.4.4   Discrete Poisson distribution

3.4.4.1   General

Consider the operation of a MEC over a period of one year:

- The number of failures of that turbine per incremental interval of X is the probability of failure in successive intervals;
- If that probability of failure, P, is constant, then the turbine operation over each interval is independent of operation in previous intervals;
- Then, the probability P has a binomial distribution with respect to the discrete random variable X.

Suppose that a constant, λ, equals the average value of failures in that month. If the variance of the failures also equals λ, then the random experiment is called a Poisson process, see Figure 3.12.

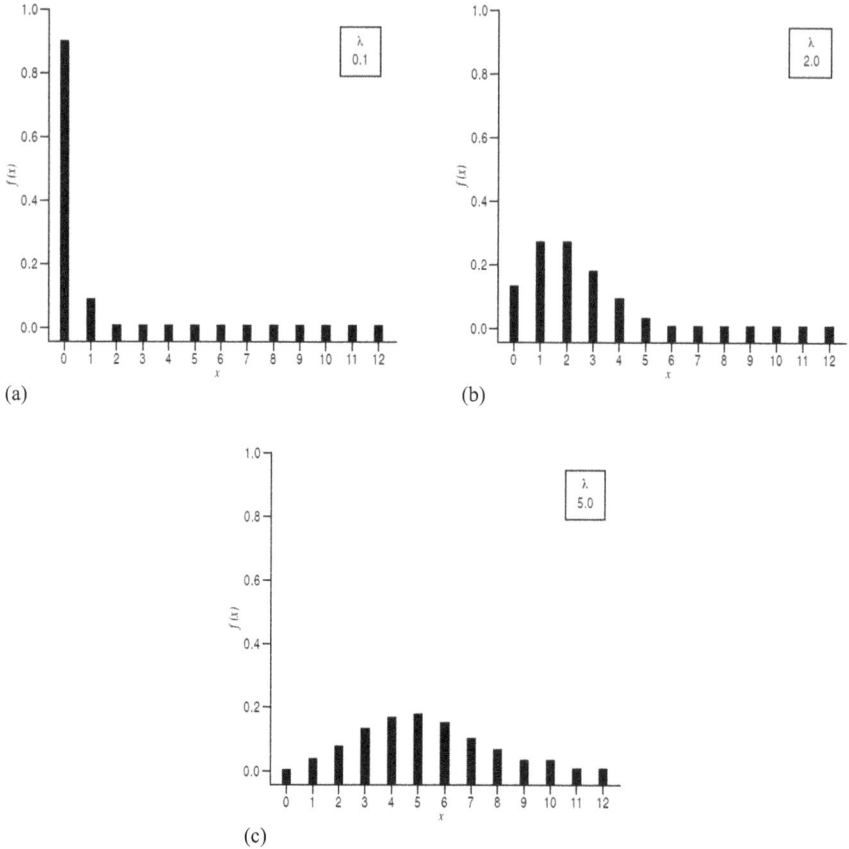

Figure 3.12   Graphical representations of Poisson distributions

Definition:

A Poisson distribution can be used as an approximation of the binomial distribution when the number of observations is large and the probability of failure is low. It can be represented by the equation below where $x = 0,1,2,3,\ldots$ is the number of failures.

$$lim_{n\to\infty}P(X = x) = e^{-\lambda}\ \lambda^x/x!, \quad \text{for } x = 0,\ 1,\ldots,\ n \tag{3.23}$$

Poisson distribution plays a special role in Reliability Theory since under broad conditions they describe the phenomenon of catastrophic failure in complex systems. The Poisson distribution defines a random variable to be a certain interval during which a number of failures occurred. The continuous variable is of interest, because as described above it could be for example calendar time, time on test, GWh produced or turbine rotations.

Let the continuous variable, x, denote the duration from any starting point until a failure is detected or in general denote the duration between successive failures in a Poisson process. If the continuous variable is calendar time, it could be censored monthly, quarterly or annually.

The starting point for measuring x doesn't matter because the probability of the number of failures in a Poisson Process depends only on the length of the interval not on the value of x.

If the mean number of failures is λ per interval, then x has an exponential distribution with parameter λ.

$$f(x) = \lambda e^{-\lambda x}, \quad \text{for } 0 \leq x < \infty \tag{3.24}$$

3.4.4.2 Point processes

A point process is a stochastic model describing the occurrence of discrete events in time or space. In reliability analysis, failures of repairable systems can be described with point processes in the calendar time domain, for example monthly, quarterly or annually, or using an operational variable, like kilometres driven or number of flying hours, which represent more closely the kind of discrete data which can be secured from the wave and tidal industry, for example failures per year.

A random variable $N_f(t)$ that represents for example the number of failure events in the interval *[0, t]* is called the counting random variable. Subsequently, the number of events in the interval *(a, b]* will be:

$$N_f(a, \ b) = N_f(b) - N_f(a) \tag{3.25}$$

The point process mean function $1\Lambda(t)$ is the expected number of failures, E, in the interval throughout time t:

$$\Lambda(t) = E[N_f(t)] \tag{3.26}$$

The rate of occurrences of failure $\mu(t)$ is the rate of change of expected number of failures

$$\mu(t) = d\Lambda(t)/dt \tag{3.27}$$

The intensity function $\lambda(t)$ is the limit of probability, P, of having one or more failures in a small interval divided by the length of the interval:

$$\lambda(t) = lim/\Delta t \rightarrow 0P(N_f(t, t + \Delta t]) \geq 1/\Delta t \tag{3.28}$$

If the probability of simultaneous failures is zero, which occur only where the mean function $1\Lambda(t)$ is not discontinuous, then:

$$\lambda(t) = \mu(t) \tag{3.29}$$

3.4.4.3   Non-homogeneous Poisson process

Assuming minimal repair, that is failed sub-assemblies are brought back to the same condition as just before the failure, the non-homogeneous Poisson process (NHPP) can be used to describe changes in reliability of repairable systems (Reliability Information Analysis Center 2010). A counting process $N_f(t)$, that is the cumulative number of failures after operational or calendar time t, is a Poisson process if:

$$N_f(0) = 0 \qquad (3.30)$$

For any $a < b \le c < d$, the random variables $N(a,b)$ and $N(c,d)$ are independent. This is known as the independent increment property. There is an intensity function λ such that:

$$\lambda(t) = lim/\Delta t \to 0(P(N_f(t, t + \Delta t)) = 1)/\Delta t \qquad (3.31)$$

Note that if λ is constant, then the process is a homogeneous Poisson process (HPP); simultaneous failures are not possible

$$lim/\Delta t \to 0(P(N_f(t, t + \Delta t)) \ge 2)/\Delta t = 0 \qquad (3.32)$$

The main property of an NHPP is that the number of failures $N(a,b)$ in the interval (a,b) is a random variable having a Poisson distribution with a mean:

$$\Lambda(a, b) = E[N_f(a, b]] = a\lambda(t)dt \qquad (3.33)$$

3.4.4.4   Power law process

An NHPP is called a power law process (PLP) if the cumulative number of failures through time t, $N_f(t)$ is given by:

$$N_f(t) = \rho t^{\beta} \qquad (3.34)$$

Therefore, the expected number of failures for a specific time interval $[t_1, t_2]$ will be:

$$N_f[t_1, t_2] = N_f(t2) - N_f(t1) = \rho(t_2^{\beta} - t_1^{\beta}) \qquad (3.35)$$

The failure intensity function, $\lambda(t)$, is then:

$$\lambda(t) = dN_f(t)/dt = \rho\beta(t_2^{\beta} - t_1^{\beta}) = \rho\beta t^{(\beta-1)} \qquad (3.36)$$

3.4.4.5   Total time on test

Practical failure data is more easily collected and managed during calendar time, t_C. Failures can then be grouped monthly, quarterly or annually as they were collected, so-called censored data. It has been shown that it is better to collect data over larger intervals, yearly or quarterly rather than monthly, as this ensures a larger population in each censored period and prevents periods being discarded for lack of data.

However, true understanding must come from knowing the actual time a part or machine has been in service. This is total time on test (*TTT*), which refers to the period from when a part is put into service to the time when it fails.

The variable *t* that appears in the various equations of the Crow-AMSAA model, see Spinato (2008), represents the time to a point process but differs from calendar time, t_C, as reported for example in MEC failure tables reported later in the book.

Reliability growth, as well as other reliability analysis, is normally carried out on the basis of specific tests made on sub-assemblies under investigation. For a repairable system, the test is stopped after a failure or a planned inspection and the number of running hours elapsed since the previous failure is recorded. After a number of failures have been accumulated, failure data are interpolated with a mathematical model, like Crow-AMSAA, to verify the achieved reliability or, using the terminology of the military standard, the 'demonstrated reliability'. The independent variable *t* of the plot is the cumulative quantity called the *TTT* that is the integral of the number of running hours of the entire population for the observed period. In this way, the hours of inactivity are excluded from the *TTT* evaluation. Using *TTT* rather than calendar time presents advantages and disadvantages and the meaning of *TTT*, for device failure data, must be clarified. First, it is in the nature of reliability engineering to deal with running hours rather than calendar time. This distinguishes reliability analysis from availability analysis. In this case, the age of many electromechanical systems can be measured by the number of cycles completed or the total running hours and this often differs substantially from the calendar age. Nevertheless, calendar time can play an important role where chemical–physical properties deteriorate with time, for example the insulating properties of a dielectric. For renewable energy WT data sets, the *TTT* in a certain interval *i*, $\Delta TTTi$ can be calculated by multiplying the number of devices, *Ni*, by the number of hours in the interval, *hi*. The recorded total hours lost from device production, *li*, in that interval are then subtracted, when this information is available. In the WT surveys, this data only included out-of-service time, rather than time when the device was unable to operate for lack of resource. The aggregated *TTT* up to an arbitrary time cell *k*, *tk*, is then

$$tk = \Sigma_{i=1}\Delta TTTi = \Sigma_{i=1}Ni \ (hi - li) \tag{3.37}$$

To calculate the *TTT* for MEC data, three considerations are necessary.

For each time interval, the surveyed devices were considered representative of the entire population. Therefore, the sample reliability for each time interval was assumed to represent the reliability of the entire population. This hypothesis was necessary to overcome one of the major data deficiencies, the variable number of devices in each time interval. In reality, any reliability improvement or deterioration spreads throughout the population with a certain rate, indicated by the shape parameter, β, as long as sample devices are assumed randomly chosen from the entire population and the usage of each device in the population is similar.

Using *TTT* has the effect of stretching the curve on the abscissa but since TTT depends on the number of devices considered, it has no absolute meaning, as calendar time would have. The abscissa *t* has significance only for the device population being examined; however, by showing the cursor at the right of Figure 3.17, calendar time can be inferred.

As the intensity function interpolates data on *TTT* rather than t_C, the fit produced is intrinsically weighted by the number of turbines in each period. A larger number of devices results in a larger *TTT* interval and the fit constraint is stronger. When *TTT* is used rather than calendar time, the abscissa stretches to a longer interval for more devices surveyed and the scale parameter increases. In cases of early or constant failures, the most important result is the demonstrated reliability, as shown in Figure 3.13.

3.4.5   *Continuous exponential distribution*

Exponential distribution plays a key role in practical computations. In many cases, the interval between two successive failures in a complex system, such as a wind, wave or tidal device, obeys an exponential distribution, see Figure 3.14.

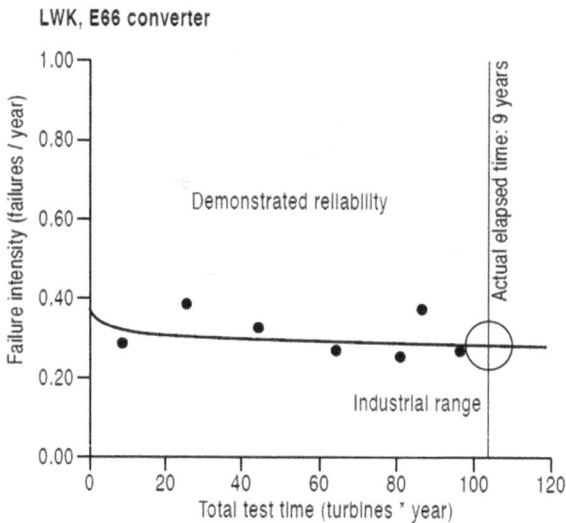

Figure 3.13   Presentation of failure intensity using total time on test, TTT, showing demonstrated reliability for a sub-assembly with some early failures, from Spinato et al. (2009)

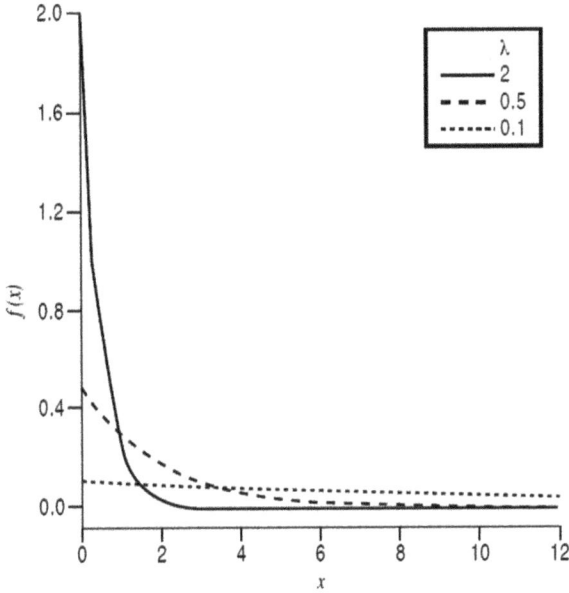

Figure 3.14 Probability density function of an exponential random variable for selected values of λ

Definition:

The random variable X that equals the distance between successive counts of a Poisson Process with mean $\lambda > 0$ has an exponential distribution with parameter, λ. Therefore, the probability density function of X is:

$$f(X) = \lambda e^{-\lambda x}, \text{ for } 0 \leq x < \infty \qquad (3.38)$$

3.4.6 Continuous Weibull distribution and its use

The Weibull distribution, see Figure 3.15, can be used to model the time to failure of many different physical systems. The parameters in the distribution provide a great deal of flexibility to model systems in which:

- Number of failures increases with time, for example bearing wear or thermal ageing.
- Number of failures decreases with time, for example early failures.
- Number of failures remains constant with time, for example random failures at the bottom of the bath-tub, caused for example by random external shocks to the system.

Definition:

The random variable X with probability density function $f(X)$ has a Weibull distribution with scale parameter $\delta > 0$ and shape parameter $\beta > 0$ given by:

$$f(x) = (\beta/\delta) \ (x/\delta)^{\beta-1} e^{-(\xi/\delta)\beta}, \text{ for } 0 < x \qquad (3.39)$$

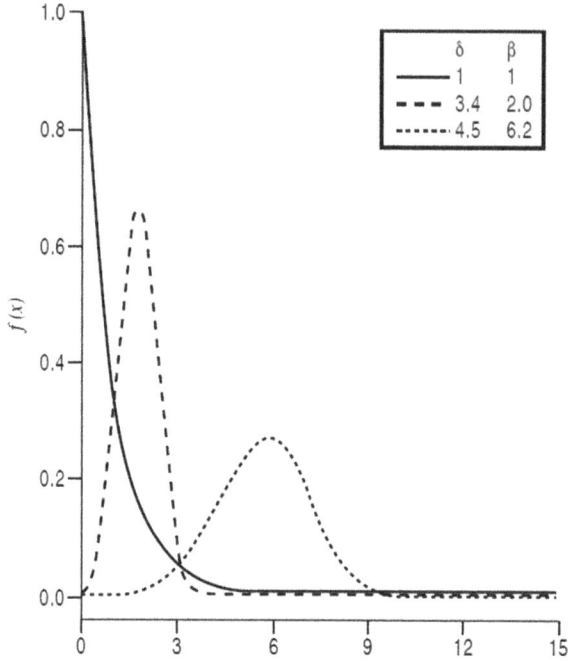

Figure 3.15   Weibull probability density functions for selected values of δ and β

The Weibull function above can be expressed in terms of time $t = x$ as follows:

$$f(t) = \frac{\beta}{\theta}\left(\frac{t}{\theta}\right)^{\beta-1} e^{-\left(\frac{t}{\theta}\right)^{\beta}} \tag{3.40}$$

The hazard function or instantaneous failure rate, $\lambda(t)$, of a component in a population of components can be obtained from $f(t)$, (3.18), because:

$$\lambda(t) = f(t)/R(t) \tag{3.41}$$

where $R(t)$ is the reliability or survivor function of a population of components, which for an Weibull distribution would be given by:

$$R(t) = e^{-(t/\theta)\beta} \tag{3.42}$$

Therefore, using the Weibull function:

$$\lambda(t) = (\beta/\theta).(t/\theta)^{\beta-1} \tag{3.43}$$

which can be simplified to:

$$\lambda(t) = \rho\beta t^{\beta-1} \tag{3.44}$$

where:

ρ is a scale parameter equal to $1/\theta^{\beta}$ and reduces to $1/\theta = \lambda$ when $\beta = 1$, which has the units year^{-1},

β is a shape parameter, which is dimensionless;

θ is the *MTTF* of the component and $\approx MTBF$, which could have the units of years.

This expression is a very powerful mathematical tool for understanding the behaviour of components, sub-assemblies and complete machines from a reliability point of view.

Figure 3.16 shows the effect on a component probability of failure, represented by a Weibull distribution, for different values of β, in this case life is expressed in arbitrary units.

So for example the curve with $\beta = 0.5$ could describe the behaviour of a population of rolling element bearings, cf Figure 3.2(b), whose probability of failure increases with time to a peak at the expected MTBF of the component.

On the other hand, the curve with $\beta = 1.5$ could describe the behaviour of an insulation component subjected to increased degradation with time and temperature and $\beta = 2.0$ could describe the progressive wear of a gearbox prematurely ageing.

Figure 3.17 shows the effect on the components hazard rates, represented by a Weibull distribution, for different values of β, against life again expressed in arbitrary units using (3.44).

When components are assembled into sub-assemblies and thence into a complex piece of machinery, such as a MEC, the aggregated component failure mode sequences results in a composite failure intensity curve, such as Figure 3.7(d), derived from (3.44), for different values of shape parameter, β.

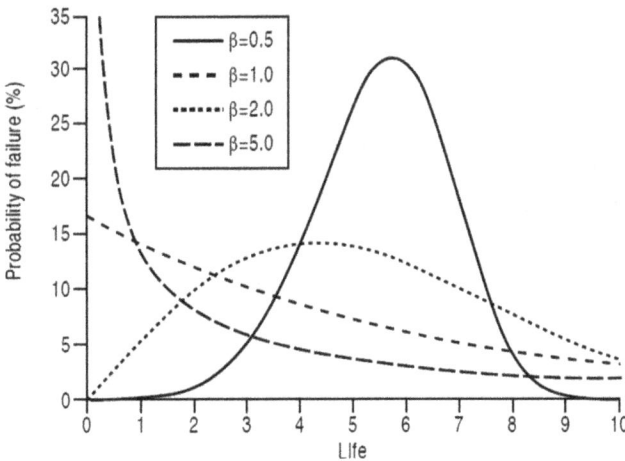

Figure 3.16   *Variation in a Weibull probability density function f(t) representation of components showing effect of different degradation processes represented by varying β*

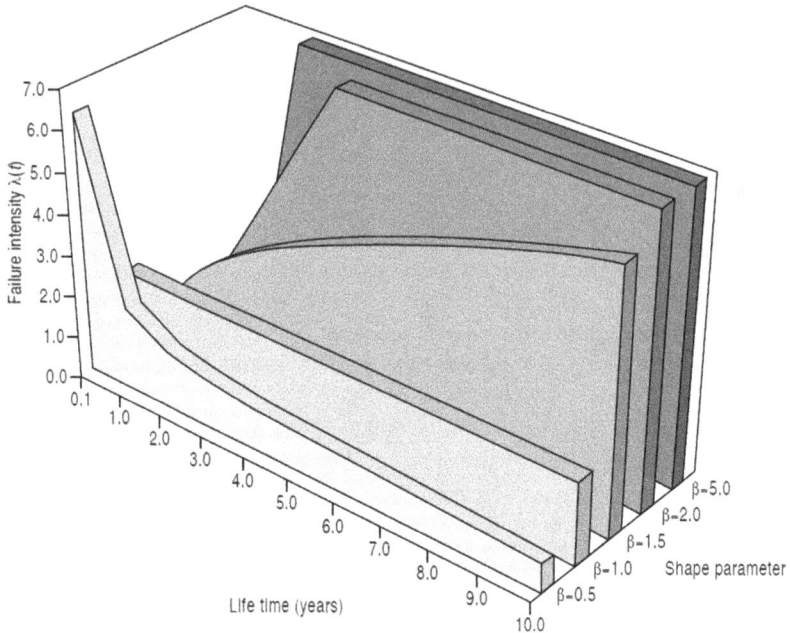

Figure 3.17   Variation in a Weibull hazard function, λ(t), of failures of a population owing to the effect of different degradation processes

So Figure 3.7(d) is a combination of three curves representing three stages of operation of the population, as considered in Table 3.1:

- Early life ($\beta < 1$): This stage is also known as infant mortality, burn-in or early failures, where the failure rate falls as the components prone to early failure are eliminated.
- Useful life ($\beta = 1$): This stage represents normal operation of the system, when the intrinsic failure rate is sensibly constant over an extended period of time, sometimes called memoryless failures.
- Wear-out ($\beta > 1$): This stage represents increasing failure rate of the system, occurring towards the end of its useful life, due to deterioration of the individual components that start to fail in increasing numbers.

However, pieces of real machinery do not necessarily follow a typical bath-tub form. The author has noted how small complex electronic equipment often does not exhibit infant mortality symptoms, because such sub-assemblies are subjected to automatic and accelerated life testing on completion of manufacture, before delivery, specifically designed to minimise early life failures.

In another way, electrical insulation systems do not tend to show a low, constant, mid-life failure rate but a rather gradual worsening in failure rate over the whole life, as the insulation material steadily deteriorates with ageing, as shown in Figure 3.16 with $\beta = 1.5$.

A similar situation occurs for mechanical systems, such as gearboxes, which steadily wear in service and this has been demonstrated in field results.

The bath-tub hazard function is instructive for those engaged in monitoring and maintenance because it demonstrates the different phases of machinery life and the character of failure modes, which we want to detect.

The low, constant, mid-life failure rate, at the bottom of the bath-tub, is achieved partly by good design and manufacture of machinery components but its base value can be reduced and duration extended by maintenance and in turn by monitoring.

So the shape of the bath-tub for a complex, repairable engineering plant is dependent upon the maintenance and monitoring regime that is adopted.

Figure 3.18 shows some examples of practical hazard functions, $\lambda(t)$, for large WT components taken from Spinato (2008) and Spinato *et al.* (2009), whose shape parameter, β, can be related to their failure mechanisms:

- For the generator $\beta < 1$: early failures due to poor manufacturing testing but good intrinsic reliability;
- For the gearbox $1.5 > \beta > 1$: steady wear through life;
- For the converter $\beta < 1$: early failures due to poor manufacturing testing and poor intrinsic reliability due to complex architecture.

To track the changes of reliability with time during the different operational phases of a product, reliability growth models have been developed most notably using the Crow-AMSAA model, see Spinato (2008). The same model can be applied to failure data collected from the field to investigate whether product reliability stays constant, or shows an improvement or deterioration with time.

3.4.7   The bath-tub curve

One can compare the failure density function, $f(t)$, derived from example data, Figure 3.7(c), with a theoretical form, Figure 3.6, of different shape, but the $R(t)$ and $Q(t)$ values, Figure 3.7(a) and (b), can be extracted for any time by integrating under Figure 3.7(c) to the left or right of that time, respectively.

The curve, Figure 3.7(d) derived from Figure 3.7(c) as the hazard rate, $\lambda(t)$, for the example data shows how the unreliability of non-repairable and repairable systems can be modelled. This so-called bath-tub curve (Rigdon and Basu 2000) represents the three different phases of a population life shown in Figure 3.19.

One of the advantages of using the PLP model for repairable systems is that its intensity function (3.44) is flexible enough to represent separately the three different phases of the bath-tub curve, Figure 3.19, based on shape parameter, β, values described in Table 3.2.

The off-shore MEC farm described in Table 3.1 and Figure 3.7 showed that at the bottom of that hypothetical bath-tub λ was about 0.09 failures/MEC/yr, far better than achieved on current UK off-shore wind farms, despite:

- The large number of failures accumulating in the second column of Table 3.1.

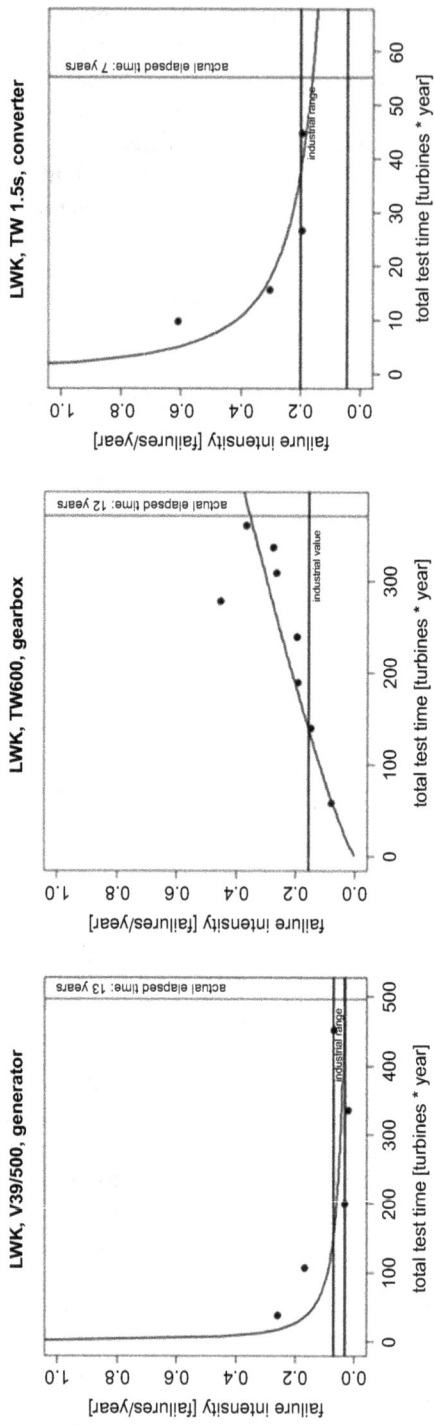

Figure 3.18   Practical hazard function, λ(t), examples for three major WT components

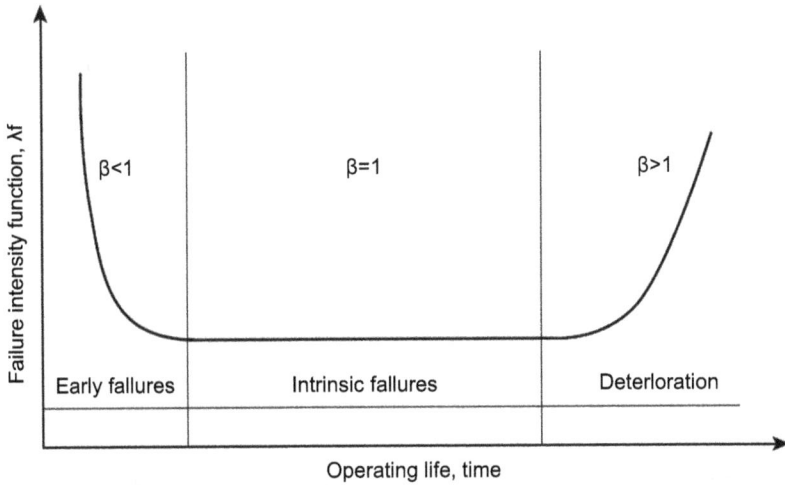

Figure 3.19   The 'bath-tub curve' for the intensity function showing how the reliability varies throughout the life of repairable machinery

Table 3.2   Values of β for different failure intensities

Value of β	Failure intensity	Reason	Model type
$\beta < 1$	decreasing with time design	improvements/alterations on field	NHPP
$\beta = 1$	constant with time $\lambda(t) = \rho$	no major design modifications – wear and tear not apparent yet	HPP
$\beta > 1$	increasing with time normal	deterioration of materials/accumulated stresses	NHPP

- The fact that current off-shore wind farms incorporate repairable devices, so that when a fault occurs a WT is swiftly returned to the operational state by effective maintenance.

Premature serial failures (PSF) have also been encountered in European off-shore wind farms, due to faulty generators, gearboxes and converters. An important modification to the bath-tub curve was therefore proposed from an industrial source, Stiesdahl *et al.* (2005), to take account of PSFs, see Figure 3.20, where a PSF distribution has been added to the bath-tub curve.

But note from Figure 3.20:

- The bottom of this WT bath-tub indicates a failure rate of 0.01 failures/WT/ yr, many times better than our hypothetical example, Figure 3.7, and better than any European off-shore wind farm, see for example Figure 3.8(a) in

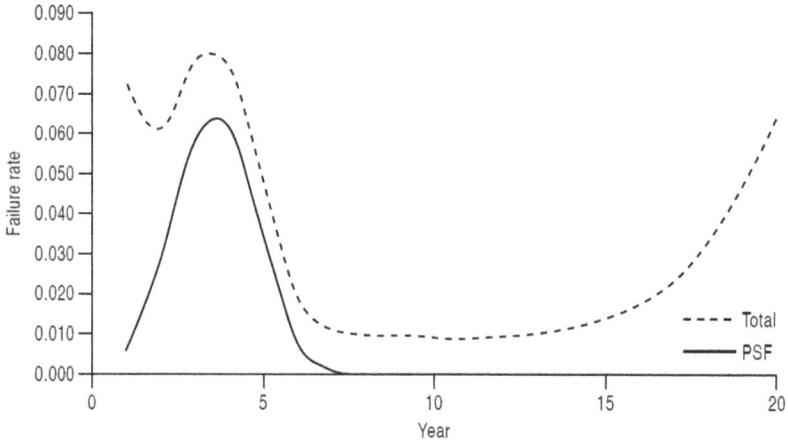

Figure 3.20   A WT 'bath-tub curve' incorporating premature serial failure (PSF)
of some sub-assemblies (Stiesdahl et al. 2005)

2005 showing 1.00 failures/WT/yr, for less arduous on-shore WT
conditions;

• Current measurements, Tavner (2012), suggest 4.00−5.00 failures/WT/yr for
off-shore WTs, although this does depend on the failure definition and outage
length;

• Safety practice in other industries, for example rail or air travel, could not
countenance the early failures shown in Figure 3.20, from PSF or any other
cause. These would be eliminated by thorough pre-testing and early operational maintenance management, both lacking in the renewable energy
industry.

3.5   Basic reliability modelling concepts for MECs

3.5.1   General

The basis for modelling the reliability of complex marine energy converters
(MECs) lies in applying the failure distributions, described in Section 3.4, to MEC
taxonomy, described in principle in Chapter 6, but the number of special conditions
needs to be taken into account for MECs associated with:

• The lack of operational data currently available;
• The fact that MECs incorporate both machinery and structural elements;
• The fact that different parts of a MEC will be exposed to different environ-
mental conditions.

All these aspects will be considered in this section to produce a concept for
reliability modelling MECs.

3.5.2   Reliability analysis, machinery vs structure vs taxonomy

The probability density curve for a MEC machinery component is likely to be a normal distribution with a life expectancy less than the expected life of the whole MEC.

The probability density curve for a MEC structural component will be a normal distribution which extends way beyond the expected life of the MEC.

A comparison between probability of failure of machinery and structural components is shown in Figure 3.21.

The machinery component, for example a generator, has a failure probability curve with an $MTBF_{machine}$ of about half the expected life of the MEC.

However, the structural component, for example a monopole, has an $MTBF_{structure}>$ expected MEC life.

The issue for the generator is to ensure that it can be replaced before $MTBF_{machine}$, whereas for the monopile, the issue is to ensure that the area to expected failure under the failure probability density curve is less than a planned structural failure rate, that is an analysis based upon extreme risk as described by Sorensen (2014).

These issues determine the design of the MEC and the probable reliability of the complete device. Concentration in device design on machinery components with $MTBF_{machine}<$Expected Life will lead to a high failure rate and low availability but low capital cost, whereas concentration on a device with structural components will lead to a low failure rate and high availability but high capital cost.

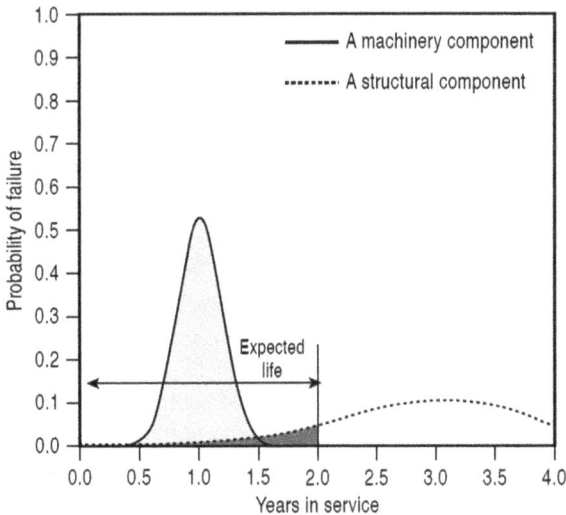

Figure 3.21   Comparing machinery and structural component continuous failure probability density functions

3.5.3   *A reliability modelling concept*

When the reliability functions defined above are assigned to a total device, its systems, sub-systems, assemblies, sub-assemblies or components with different reliability functions will be obtained. For the purposes of studying MEC technology at this stage of development, the author chooses to operate with two reliability characteristics for mathematical modelling:

- Reliability survival function $R(t)$
- Constant failure rate λ, describing probabilities of failure of non-repairable parts for comparison.

Mathematically, reliability survivor function $R(t)$ or probability that a system or part will survive after a specified time t can be expressed as:

$$R(t) = 1 - f(t) = 1 - \int_0^1 f(t)\ dt \equiv \int_1^\infty f(t)\ dt \qquad (3.45)$$

The hazard rate $\lambda(t)$ indicates the instantaneous probability that a given system or part will fail, assuming that the system/part is operational, defined as:

$$\lambda(t) = f(t)/R(t) = f(t)/(1 - f(t).R(t)) \qquad (3.46)$$

The probability is usually high at the start of operational life as a result of manufacturing defects or mishandling. Towards the end of the system's or part's life, the probability of failure increases due to general wear and tear and the system/part cannot be repaired or replaced. However, there is usually an intermediate period where the probability is more or less constant. As a function of time, this probability has the 'bath-tub' shape shown in Figure 3.4. For reliability analysis, the intermediate period where the hazard rate is constant will be used for prediction.

If the hazard rate $\lambda(t)$ becomes constant, λ, as it does during the intrinsic failures period, then:

$$f(t) = \lambda e^{\lambda t} \qquad (3.47)$$

and

$$R(t) = e^{\lambda} \qquad (3.48)$$

The reliability survivor function $R(t)$ shown in (3.45), for a period $t = 1$ year between maintenance services, will be used in this book as the reliability characteristic for MEC analysis.

For a device with all parts in series, i.e. where the failure of any one part will cause system failure if there is no redundancy, the total predicted system reliability $R(t)$ over time t is the product of the reliabilities of the component systems:

$$R(t) = \Pi_i R_i(t) \qquad (3.49)$$

Table 3.3   Reliability characteristics for different parallel non-repairable systems

Systems	Active redundancy	$R(t)$	$\lambda(t)$
1	1 of 1 must be working	$e^{-\lambda t}$	λ
2	1 of 2 must be working	$2e^{-\lambda t} - e^{-2\lambda t}$	$2\lambda/3$
3	1 of 3 must be working	$3e^{-\lambda t} - 3e^{-2\lambda t} + e^{-3\lambda t}$	$5\lambda/6$

and the total predicted system failure rate λ_{tot}, where part i has a constant failure rate λ_i, yields:

$$\lambda_{TOT} = \Sigma\lambda_i \tag{3.50}$$

Different parallel system configurations can be described as shown in Table 3.3, which shows the reliability characteristics of three different system arrangements, assuming each system is non-repairable.

A summary of effective redundancy equations for calculating reliability survivor functions and failure rate estimation for different system configurations is presented in RIAC&DACS (2005).

For non-repairable systems with independent sub-assemblies, when the hazard function or failure intensity $\lambda(t)$ is constant λ with mission time t, an exponential probability distribution is the most appropriate to describe time to failure. This distribution has been chosen by the authors due to its simplicity and also because part field failure rate data is not available and surrogate constant failure rates are used, which represent random variables. Surrogate sub-assembly' constant failure rates are related to the useful lifetime of the hazard failure rate curve, Figure 5.2, and this data is applied to TSDs and analysed in Section 5.6.

Modarres *et al.* (2010) argue that the exponential distribution is a good model for representing systems and complex, non-redundant components consisting of many interacting parts.

TSD sub-assemblies should have long random-failure regions by design. Their architectures will consist of complex, non-redundant electronic and mechanical components; therefore, it is reasonable to use the exponential time-to-failure distribution rather than other distributions. For example, a Weibull time-to-failure distribution model would require more data than are currently available for TSD sub-assemblies.

3.6   Summary

This chapter has presented the essential reliability mathematics necessary to understand how data collected from devices and device farms can be analysed and compared for reliability purposes.

It shows that simple methods can be used to extract essential information and the overall results which can be obtained. However, it also shows that care must be taken in manipulating the data to ensure that interpretations are sound.

This chapter has shown that causality must be the guiding principle when considering MEC reliability:

- First, the operator must be aware of the failure modes and root causes for the device being considered;
- Second, causality must be traced through prospective failure sequences, possible by the use of cause and effect diagrams:
  - The probability of failure of a component of a device can be described by a probability density function for that component;
  - The resultant hazard function curves for each component in a sub-assembly can be aggregated to give a prospective life curve for that sub-assembly;
  - These can then be aggregated to give the life curve of the device;
  - The aggregation of cause and effect diagrams and hazard functions needs to be done with care, taking account of the assembly configuration of the device and the reliability dependence of components;
  - From the aggregate hazard function, a model for the complete device could be derived.

- Third, MEC monitoring needs to address the root causes that have a slow failure sequence to the failure mode and this information can be derived from knowledge of the failure modes in operation in a device.

In Chapter 1, the author proposed three different courses of maintenance action:-

(i)   Breakdown maintenance, implying uncertainty and a high spares holding;
(ii) Fixed-time interval or planned maintenance, implying extended shutdown or outage periods and lower availability;
(iii) Condition-based maintenance, implying low spares holding and high availability.

Real engineering plant demands a flexible combination of the above maintenance regimes and option (i) is probably not available for MECs.

This chapter has shown that by addressing failure modes that are slow to mature, monitoring could significantly affect the detection of faults before they occur. Monitoring can then form part of planning fixed-time interval maintenance (ii) and will be at the heart of maintenance for method (iii).

Prior to determining the type of equipment and frequency of monitoring, some consideration should also be given to whether the cost of implementing such a program is justified. Factors involved in this decision may include:

- Replacement or repair cost of the equipment to be monitored;
- Criticality of the plant to safe and reliable operation;
- Long-term future of the facility in which the equipment is installed.

Assuming that consideration of the above points indicates that the expenditure is justified, the following points will aid in determining how the condition monitoring program should be implemented.

- Design;
- Manufacture;
- Installation;
- Operation;
- Maintenance.

In the majority of cases, the end users of devices are dealing with existing plant and will have to consider all five of the points identified above. This task is complex because often there is very little design information, the manufacturer may no longer exist and installation, operation and maintenance records may be incomplete. The paucity of information can lead to the assumption that the age of the equipment indicates the need for more rigorous monitoring. However, this is not always the case since the refinement of design tools and the constant pressure on manufacturers to reduce costs has, in some cases, led to decreased design margins.

The practicability of these approaches for devices depends upon the application of the device and the engineering plant it serves. That is the basis of this book.

Chapter 4

Reliability prediction method for wave and tidal devices

4.1 Introduction

It is proposed that a reliability prediction model for prototype MECs could predict incident failures, analyse a system's future performance and compare different architectures, so that lower failure rate devices could be developed. Challenges in establishing such a method include an absence of historical data and the low level of deployment of wave and tidal stream generation technology in comparable locations.

In speaking of MEC reliability concepts, it is useful to present the definition of probability-based reliability and reliability modelling: based on MIL-STD-721C, reliability in general is the probability that a part or system will perform its intended function for a specific time interval under stated conditions. It is measured quantitatively and consists of several reliability characteristics, the results being different for non-repairable and repairable systems/sub-systems/assemblies/sub-assemblies/components.

This chapter sets out a system assessment method to quantitatively evaluate MEC technology reliability, based on the theory in Chapter 3, and the device taxonomies in Chapter 6.

Such a method could then be used to analyse the future performance of devices, and compare different architectures, so that lower failure rate devices can be developed and deployed.

4.2 Different reliability prediction and assessment methods

4.2.1 General

Reliability prediction and assessment can be defined as the process of quantitatively assessing a system design, relative to its specified reliability. Prediction analysis applies appropriate models, failure rates and repair rates to evaluate systems, sub-systems, sub-assemblies or component reliability parameters.

In speaking about types of reliability prediction and assessment methods, it is important to understand that there are currently two schools of data analysis method, with two approaches to the meaning of probability and applications of probability to different scenarios. These are:

- The classical approach, used in this book;
- The Bayesian or subjectivist approach, described in Section 4.2.6.

These two approaches can be applied to MEC engineering reliability prediction and assessment analysis, but at different stages of design, development and maintenance. The type of data to be used will depend upon the chosen approach. This author intends to apply the classical rather than the Bayesian approach, because it was developed in MIL-HBDK217 (1991), RIAC&DACS (2005), RIAC (2010), Modarres *et al.* (2010) and is well proven in practice.

The different models of the classical approach will be described, as this will highlight the decision-making process for the MEC reliability prediction modelling method proposed, as described in Section 4.3.

Table 4.1 provides an overview of the most commonly used methods to evaluate inherent device reliability. It must be accepted that different reliability models are subject to different assumptions, based on different reliability data. It is noted that MEC developers consider in-service device performance as based upon time, usually at least one year in-service, without repair. MECs are also subject to ageing and depreciation, affected by different operational environments.

4.2.2   Reliability modelling and prediction (RMP)

As defined by RIAC&DACS (2005), a reliability model is a visual representation of the functional interdependencies of a system, with a framework of prediction analysis for reliability estimates, which will guide design decisions. Derived models assist in device failure predictions, visual representations of series, and parallel configurations and redundancies; they also show the reliability characteristics and factors possible for systems failures.

Models are derived from functional requirements defined by a functional block diagram (FBD) and provide a basis for the development of RBDs for calculating the total system annual failure rate or total system reliability for the devices. The RBD is used primarily to quantify the reliability survivor function of a system, subsystem or total device, thus can be called an assessment or prediction method. Models can be simple or complex and may include variation in environments, operations, controls and human interactions. Development of the models depends on the type and amount of reliability data available and the criticality of the system(s) being analysed. Simple systems RBD modelling are shown in Figures 2.20, 3.9 and 3.10. Each block may represent the maximum number of parts with assigned λ_i under a specific environment. Examples of derived graphical models for TSDs are presented in Chapter 5 and Appendix 3. According to RIAC&DACS (2005), even a simple model will help guide design decisions to improve overall reliability. Appropriate judgment must be used.

Table 4.1 Most commonly used system reliability assessment methods

Methods	Purpose	Application	Limitations	Use
Reliability modelling and predictions (RMP)	Quantitatively evaluates reliability of competing designs and direct reliability-related design decisions Modelling uses FBD and RBD Prediction and Parts Count Reliability Prediction	Perform early in the design phase. More beneficial for newly designed hardware. Applicable to all types of hardware Surrogate data can be used before part-level testing provides first-hand data. Predictions using surrogate data to be used for comparing designs, not for making absolute predictions	Conservative deterministic assessment	Proposed for wave and tidal device reliability assessment
Failure modes and effects analysis (FMEA)	Bottom-up approach to identify single failure points and their effects Assists in the efficient design of built-in and fault isolation tests. Identifies interface problems	Perform early in the design phase to help improve design. Use when investigation of all possible failure modes is critical. More appropriate for equipment performing critical functions	Labour-intensive with complex or interconnected paths Does not identify potential failures due to 'human error'. Need more data than an RBD	Will not be used for wave and tidal device reliability assessment, as information for an FMEA is not available
Fault tree analysis (FTA)	Top down functional analysis to identify effects of faults on system performance. Systematic deductive method for defining a single specific undesirable event. Deterministic assessment FTA can be considered an assessment but only if failure rates or probability of occurrence can be	Use during initial device design when primary concern is safety, human error or some other explicit 'top event'. More limited in scope and easier to understand than FMEA. Results may be useful for trouble-shooting after the device is built	Difficulty in distinguishing between dependent and independent events in the construction of the fault tree	Will not be used for wave and tidal device reliability assessment, as information for FTA is not available

(Continues)

Table 4.1 (*Continued*)

Methods	Purpose	Application	Limitations	Use
	assigned to all sub-systems; otherwise, only the single point failures and multiple failure sets can be identified but no estimate of reliability can be made			
System simulations analysis: Monte Carlo algorithms: software tools – RAPTOR, BlockSim, and AvSim+	To quantitatively evaluate the reliability of competing designs. To direct reliability-related design decisions. Uses system models, failure rates and repair rates to estimate device reliability. Enables trade-off with respect to different design approaches. Used for complex topologies, when reduction to series and parallel connections not possible (i.e. RBD is insufficient). Used to evaluate additional aspects of system performance. Discrete event-driven simulation. Probabilistic reliability assessment. Provides dependency information, capacity information, changes between operating phases	Perform early in the design phase, as detailed data becomes available. More beneficial for newly designed hardware. Applicable to all types of hardware	Requires a lot of data	Proposed for future studies of seasonal variations in wave and tidal device reliability
Bayesian model	Further assessing systems reliability and uncertainties	Future prediction of sub-assemblies	Selection of a prior distribution, a lack of available data	Will not be used for wave and tidal device reliability assessment

4.2.3 Failure modes and effects analysis (FMEA)

An FMEA is used to determine the parts in a system that can fail, and what the cause of this failure is likely to be. This can be done with a quantitative method, whereby a failure mode is assigned a failure rate, and the criticality of said mode is determined in relation to other modes. The FMEA can determine which failure modes can be resolved by changing the design, and which ones might never be resolved, due to either limited resources or time, or because they are too minor. See Section 9.1.2.3. The FBD and RBD are prerequisites for performing an FMEA. The drawbacks to FMEA are:

- That occurrence probability should be measured quantitatively, which can only occur if failure rates can be assigned to failure modes. If that cannot be done, an FMEA can only qualitatively identify critical failure modes occurring with higher probability.
- An FMEA cannot identify potential failure modes due to human error. An FMEA of equipment with highly complex or interconnected paths is labour-intensive: a decision to use it should be based upon the knowledge of the device design team.

4.2.4 Fault-tree analysis (FTA)

Like an FMEA, an FTA can be quantitative or qualitative and can be used to find multi-point failures. Failure rates must be applied to failure modes to obtain quantitative results, again like an FMEA. Without which qualitative analysis would show only single point failures, then FTA scenarios can show two, three or more failures. Difficulties lie in the identification of dependent and independent events. Figure 4.1 illustrates a graphical example of a system event in the form of an FTA.

According to Smith (2001), the choice of the FTA method is subjective; a researcher chooses which method is the most appropriate, taking account of the tools available, the complexity of the system and the most appropriate graphical failure logic representation. RBD and FTA methods are deterministic, that is the device's reliability will be presented by only one factor. This means that for a derived series/parallel model, with failure rate data, only one solution is applicable. The limited availability of applicable failure rate data precludes the use of any FTA method.

4.2.5 Monte Carlo simulation

A different method to estimate system reliability was introduced in 1964, using a Monte Carlo simulation to sample random numbers from the underlying probability distributions. System failure rates are described by a probability distribution function (PDF), so the simulation proceeded by sampling these PDFs over an interval, generating uniformly distributed random numbers [0, 1]. The random numbers are then tallied for their failure results, the greater the number of trials, the closer the results should approach to a theoretical steady state. Simulation quality is

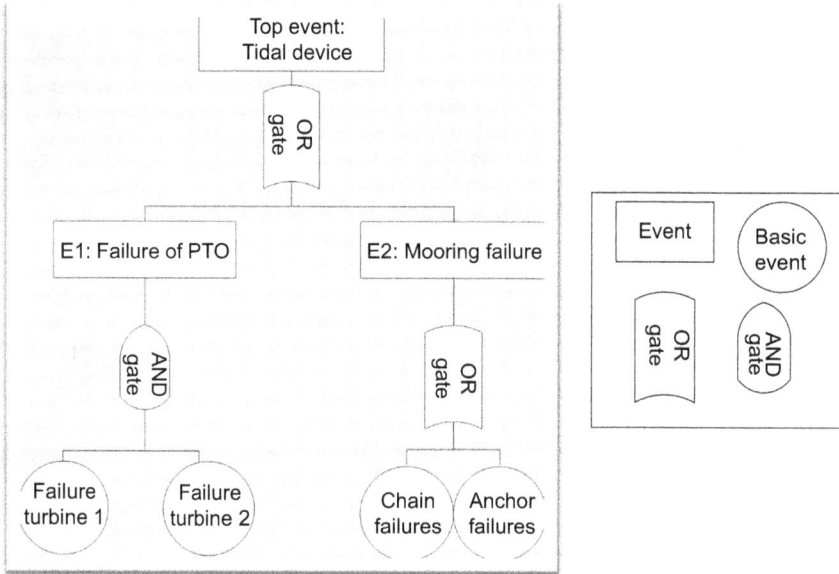

Figure 4.1   Illustrative example of fault tree analysis (FTA) for a TSD failure, Thies (2009)

determined by the quality of input data to real-life failure information, that is the PDF accuracies. Without this data, a Monte Carlo method cannot operate. Therefore, in a situation where MEC failure rate distributions are unknown, as would be the case in their early development, the method cannot be used.

Nevertheless, Monte Carlo simulation is a useful tool for assessing system reliability uncertainties when there is lack of knowledge about the true failure rate (λ) data, making it extremely useful for assessing a system's reliability uncertainties. Lack of a true λ is compensated for by substituting, subjectively, a probability distribution to predict λ. By estimating λ graphically, one can use it to compensate for the unknown λ.

4.2.6   Bayesian subjective modelling

Bayesian modelling for reliability prediction is based on subjective interpretation of analysed data, where $P(E)$ is a measure of the belief one holds in a specified event E (Modarres *et al.* 2010).

In Delorm (2014), the applicability of the Bayesian approach was demonstrated for the reliability of a main bearing of a TSD, where the method used part improvement simulation analysis.

However, the method had limitations because it required complex analysis of detailed bearing design information, unavailable in the pre-design stage. So this would be unsuitable for device architecture comparisons at the early design stage.

The Bayesian method is, however, a useful tool for assessing further systems reliability uncertainties and future prediction of incident failure of parts, when the designers do have some information but not much data. The major problem is selection of the prior distribution, which depends upon the amount of available data and its format. This is another approach for analysing MEC sub-assembly uncertainties, which could be used as more reliability data become available.

4.3 Proposed RMP method

4.3.1 Reliability modelling and prediction

RMP is applicable to both the mechanical and electrical sub-assemblies incorporated in WECs and TSDs. Based on the work by YARD (1980), AME (1992) and RIAC (2010), the sequence used for RMP predictions should be as follows:

- Perform a parts classification for each device, assigning codes for each sub-assembly using robust methods;
- Establish a schematic diagram for each device, based on the defined configuration;
- Derive an FBD from that schematic diagram, showing the logical and functional interdependencies between sub-systems, assemblies and sub-assemblies, constituting an RBD;
- In the absence of historical reliability data, collect reliability data from surrogate data sources, using them to allocate failure rates for each FBD sub-assembly;
- In the unknown environment without historical reliability data, establish lower and upper bound failure rates, λ_{Gi_min} and λ_{Gi_max}, for each sub-assembly from surrogate data, and use the upper bound as the more conservative value;
- Adjust surrogate failure rate data to the tidal environment using two extreme failure rate estimates, upper and lower bounds;
- Calculate predicted tidal environment failure rates;
- Evaluate the total device reliability, using the PCRPT of MIL-HDBK-217F (1991), MIL-HDBK-338B (1998), assuming that sub-assembly times-to-failure were exponential, that is hazard rates are the result of random failures, and a constant failure rate applies.

The parts classification of the systems and sub-systems was performed according to the international VGB Reference Designation System, set out for WTs in VGB PowerTech (2007), because the taxonomy of wave and tidal devices is similar to WTs. System reliability considerations require identification of all the sub-systems of any system. The criticality of each part may not be identical but the failure rates are statistically significant.

The FBDs present a clear picture of the functional interdependencies for each device. By assuming that each sub-assembly is independent of others and that sub-assemblies operate in a single environment, the overall tidal device reliability, based on WT experience, can be analysed as a series or parallel network, using an RBD.

The constituted RBDs of five horizontal axis TSDs with different architectures, presented in Figure 6.4 and Appendix C, will be used primarily to quantify the reliability survivor function of each device, thus can be called an assessment or prediction method.

RIAC (2010) stated that in order to evaluate the total system reliability, the PCRPT (parts count reliability prediction technique) can be used on RBD models. This technique is based on the assumption that the average failure rate for each subsystem or part is constant during useful life, Figure 4.1, and that the time to failure of sub-systems is exponentially distributed.

4.3.2   Reliability data from surrogate sources

To analyse prospective MEC sub-assemblies, a portfolio of surrogate data (PSD) needs to be created from different industries. This is given in Section 6.4.2.

4.3.3   Reliability using environmental modification factors

The main question mark over the surrogate data approach is the relevance of these data to the device environment, given the environments from which the surrogate data came, which are summarised in Table 4.2.

Further treatment is therefore needed to apply the surrogate data to a marine environment, by the use of environmental adjustment factors. Three different operational environments, found in tidal devices, were to be applied as presented in Table 4.3:

- GB-ground benign: protected environment
- GF-ground fixed: severe environment
- GM- ground mobile: special environment
- NS-naval sheltered: normal environment
- NU-naval, unsheltered: severe environment

In general, WTs and electrical equipment are in the GF environment, whereas WECs and TSDs are in the NS or NU environment. Tidal environment

Table 4.2   Environments of surrogate data sources used in the model

Surrogate data source	Naval, unsheltered: severe environment NU	Naval, sheltered: normal environment NS	Ground, fixed: severe environment GF
LWK, WMEP	–	–	X
OREDA (1984–2015)	X	X	X
NPRD-95	X	X	X
MIL-HDBK217F	X	X	X

sub-assembly failure rates could be predicted using surrogate data by applying a GF to NU or NS adjustment factor.

Environmental adjustment factors for electrical and electronic components from MIL-HDBK 217F (1991) are tabulated in RIAC&DAC (2005). This data must be modified for this study, which considers component failure rates, λ_i, because MIL-HDBK-217F multipliers were intended for MTBFs. The modified data is presented in Table 4.3, and environmental definitions are described in references MIL-HDBK 217F (1991) and SD-18 (2006).

Applying appropriate environmental adjustment factors will reduce errors in the subsequent predictions, a previously defined rationale did not exist for this approach, so the use of two failure rate estimates is proposed as set out in Table 4.4.

Table 4.3   Environmental adjustment factors, π_{Ei}

	Standards		Environmental factors[a]				
		MIL-HDBK-217F (1991)	GB	GF	GM	NS	NU
	MIL-HDBK-217F (1991)	SD-18 (2006)	Protected	–	–	Normal	Severe
From this environment	GB	Protected	X	2.0	5.0	3.3	10.0
	GF	–	0.5	X	0.4	1.7	3.3
	GM	–	0.2	0.4	X	0.7	1.4
	NS	Normal		0.6	1.4	X	2.0
	NU	Severe	0.1	0.3	0.6	0.5	X

[a]Environments defined in standards, see definition above.

Table 4.4   Failure rate estimates

Failure rate estimate	Method using surrogate data	Limitations
Conservative FRE_{con}	No environmental adjustment applied. $\lambda_i^{(b)} = \lambda_{Gi\_max}$	Represents a conservative failure rate for a branch, b, but neglects environmental conditions.
Environmentally adjusted conservative FRE_{env}	Multiplied by an environmental factor, π_{Ei}. $\lambda_i^{(b)} = \lambda_{Gi\_max} * \pi_{Ei}$ For mechanical components: $\pi_{Ei} = 1$. For electrical/electronic components: π_{Ei} as defined in Table 4.3	Represents a conservative failure rate for a branch, b, but takes account of environment.

For example in the next chapter, surrogate data taken from OREDA (1984–2015), an oil and gas environment differing from MECs, the author will use OREDA failure rates at the upper limit of 90% confidence interval, to provide a conservative or upper bound reliability limit.

4.4   Reliability prediction model calculations

Individual sub-assembly failure rates were combined into a total failure rate for two alternative operating constraints as follows:

1. Predicted total failure rates based on the assumption of a non-repairable series assembly of independent sub-assemblies, operating up to full 100% of device power output for TSD1 to TSD5 for one calendar year.
2. The same but operating up to full 100% of device power output for TSD1 to TSD5, or 50% of device power output for TSD2 and TSD4 that incorporate twin-axis turbines and therefore have implicit redundancy.

To evaluate assembly reliabilities, the PCRPT was used and the FBD was simplified to a series model, reducing any redundant sub-assemblies to an equivalent single branch, b, then using sub-assembly failure rates to calculate the predicted failure rate for the ith generic part according to Table 4.4:
either

$$\lambda_{i\_B} = \lambda_{Gi\_max}, \text{failures/year} \tag{4.1}$$

or

$$\lambda_i^B = \lambda_{Gi\_max} * \pi_{Ei}, \text{failures/year} \tag{4.2}$$

The predicted assembly failure rate models are as follows:
Series network: For a TSD with all sub-assemblies in series, the device will fail if any one of the sub-assemblies fails. For a series reliability model of independent sub-assemblies with constant failure rates, the reliability model forms and total failure rate are expressed as:

$$\lambda\text{tot}_N\text{s} = \sum_{b=1}^{N_s} \sum_{i=1}^{N_b} \lambda_i^{(b)} \tag{4.3}$$

Series parallel network: For a device with N_s sub-assemblies in series and N_p assemblies with two identical branches in parallel with constant failure rates, the reliability model forms and total failure rate are expressed as below.

An example could be a MEC with twin or more redundant axis drive train branches, as will be encountered in Chapter 6.

This could allow up to 25% power production (DT), an uninterrupted electrical assembly (B), a redundant ancillary assembly (XA) and twin-axis nacelle configurations (U):

$$\lambda tot = \lambda tot_N s + \left(\frac{2}{3}\right)\lambda tot_N p \tag{4.4}$$

$$\lambda tot_N p = (\lambda DT + \lambda B + \lambda XA + \lambda U + \ldots) \tag{4.5}$$

$$\lambda tot = \sum_{b=1}^{N_s}\sum_{i=1}^{N_b}\lambda_i^{(b)} + \frac{2}{3}(\lambda DT + \lambda B + \lambda XA + \lambda U + \ldots) \tag{4.6}$$

where:

$$\lambda DT = \sum_{i=1}^{N_{DT}}\lambda_i^{(DT)} \tag{4.7}$$

This is the total failure rate of a drive train (DT), equal to the sum of the failure rates of the single branch DT sub-assemblies:

$$\lambda B = \sum_{i=1}^{N_B}\lambda_i^{(B)} \tag{4.8}$$

is the total failure rate of an uninterrupted electrical assembly (B), single branch twin-axis assembly

$$\lambda XA = \sum_{i=1}^{N_{XA}}\lambda_i^{(XA)} \tag{4.9}$$

is the total failure rate of an ancillary assembly (XA), single branch of twin-axis redundant assembly

$$\lambda U = \sum_{i=1}^{N_U}\lambda_i^{(U)} \tag{4.10}$$

is the failure rate of a single branch twin-axis nacelle or support structure (U).

Therefore, the device reliability survivor function $R(t)$ can be calculated as follows:

For series network

$$R(t) = e^{-\lambda totNst} \tag{4.11}$$

For series/parallel network

$$R(t) = e^{-\lambda tot} \tag{4.12}$$

The model uses a standard formula for assemblies with twin or more axis redundant drive train branches, as is encountered in some MECs, see Chapter 6, to represent a single reliability parameter curve. The method can be extended to more complex assemblies using standard reliability equations published in RIAC&DACS (2005).

4.5   Summary

As MEC device design and deployment is at an early stage of development, there is limited availability of reliability data.

This makes the use of classical probabilistic FMEA, FTA reliability models, valid Monte Carlo or Bayesian simulations almost impossible. However, a valid reliability assessment method is available, that is RMP.

RMP adopts a combination of:

• Graphical models consisting of FBDs and RBDs, presented later;
• Mathematical models based on PCRPTs;
• A portfolio of surrogate data (PSD), described below.

Where data is lacking, surrogate data sources, PSD, can be identified and adjusted to the marine environment, for use in MEC models. PSD draws on the following sources adjusted to the marine environment:

• WT Databases: Hahn *et al.* (2007), Spinato *et al.* (2009), Tavner (2012);
• Marine Database: OREDA (1984–2015);
• Generic Reliability Databases MIL-HDBK-217F (1991), NRPD-95 (1995);
• Other Integrated Reliability Databases, including IEEE, Gold Book (1997).

This chapter, therefore, proposes RMP using PSD, based on MIL-HDBK-217 and RIAC (2010) methods, as a reliability assessment method for MECs at the conceptual design stage. A simplified version of that reported in Delorm *et al.* (2011) and presented in Delorm (2014) can be applied to generic MEC designs, incorporating known main features, so that a view can be taken of different designs on the impact of reliability.

The method proposed in this chapter will now be applied in Chapter 6 to three WECs and three TSDs of generic manufacture rated, at 1–2 MW on a system/sub-system/assembly/sub-assembly basis to analyse and compare devices.

The method could also be applied by the design developers of specific devices, who will have access to the detailed design information necessary to define more complete reliability models.

Chapter 5
Practical wave and tidal device reliability

5.1 Introduction and reliability data collection

No international standard for the collection of reliability data from Marine Energy Converters (MECs), such as wave and tidal energy devices, has been established yet. However, this chapter proposes a standard methodology for improving the reliability of Wave and Tidal Generation Devices. However, the following two projects from wind power reliability will be relevant to MECs:

- An EU Project, ReliaWind, set out a standardised method for reliability data collection for wind farms, including off-shore, which is described by Tavner (2012) and included as an appendix to this book, Appendix E.
- IEA Wind Task 33 (2016) is developing a methodology, using the above as a starting point, that will be applicable to off-shore wind and therefore to MECs.

This chapter sets out what we know about the reliability of the components we will encounter in the types of MEC to be analysed.

5.2 Typical root causes and failure modes

5.2.1 General

It is important to distinguish between root causes, which initiate the failure sequence and can be detected by monitoring systems, and failure modes, which terminate the failure sequence.

After a failure, operators are used to tracing the sequence back from the failure mode to the root cause, in order to determine the true cause of failure. That is the process of root cause analysis (RCA), or asking why a failure occurred.

On the other hand, the designer of a system, to increase reliability and raise availability, must keep in mind the need to predict the reverse of that process, tracing how a failure develops, as shown in Figure 3.1.

On this basis, the author proposes following the most common root causes and failure modes in MECs, based on the descriptions of Chapter 3 and Appendix E. They are similar to the root causes and failure modes identified by the IEEE, Gold Book (1997). It is surprising how few root causes and failure modes there are and the reader should note that these failure modes and root causes are generic and could be applied to many different sub-assemblies and components of the device.

5.2.2   Root causes

- Defective design or manufacture
- Defective material or component
- Defective installation
- Defective maintenance or operation
- Ambient conditions
- Over-speed
- Over-load
- Low-cycle fatigue or shock load
- High-cycle fatigue or excessive vibration
- Component failure
- Excessive temperature
  - Hydraulic power pack
  - Electrical windings
  - Mechanical gears and bearings
- Excessive dielectric stress, steady or transient
- Debris or dirt
- Corrosion

5.2.3   Failure modes

- Electrical
  - Software failure
  - Electrical integrity, including insulation, failure
  - Electrical trip
- Mechanical
  - Shaft, bearing or gear failure
  - Lubrication or hydraulic power pack failure
  - Cooling failure
  - Mechanical integrity failure

5.3   Current reliability knowledge

5.3.1   *Reliability analysis and industrial reliability data for sub-assemblies*

There is currently almost no recorded reliability data from operating MECs, so the approach of this section will be to put reliability into context by considering the mean time between failures (MTBFs) of some typical parts of MECs, where that information is known.

Such data can be notoriously difficult to find but some information, particularly about electrical equipment MTBFs, is available from reliability surveys described below.

5.3.1.1 Wind industry databases: WMEP, LWK and Windstats

Hahn *et al.* (2007) analysed the reliability of German WTs and parts, following on from Schmid *et al.* (1991), providing figures of failure frequency and downtimes voluntarily reported to ISET over 15 years and evaluated under Germany's Scientific Measurement and Evaluation Programme (WMEP) '250 MW Wind'. This database was for repairable, on-shore WTs, documenting 60,000 maintenance and repair episodes in which averaged annual evaluations show WT availability in the 97%–98% range.

Data from 6,000 WTs in Germany and Denmark over 11 years of operation were surveyed by Spinato *et al.* (2009), focusing on a sub-set of 650 off-shore machines, from which the data about the reliability of generators, gear-boxes and electricity converter sub-assemblies were analysed. The authors concluded that, although the reliability is considerably below that of such sub-assemblies in other industries, reliability was improving with time.

European on-shore WT database data are summarised in Figures 5.1 and 5.2.

5.3.1.2 Oil and gas industry database: OREDA

Since 1983, OREDA has collected reliability data from a wide range of equipment used in oil and gas exploration/production, based on the experience of oil companies operating in the North Sea and the Adriatic Sea (OREDA, 1984–2015). Most off-shore and subsea equipment are covered by this database. Failure-rate data is presented in a time window of 2–4 years of operation for off-shore equipment, and

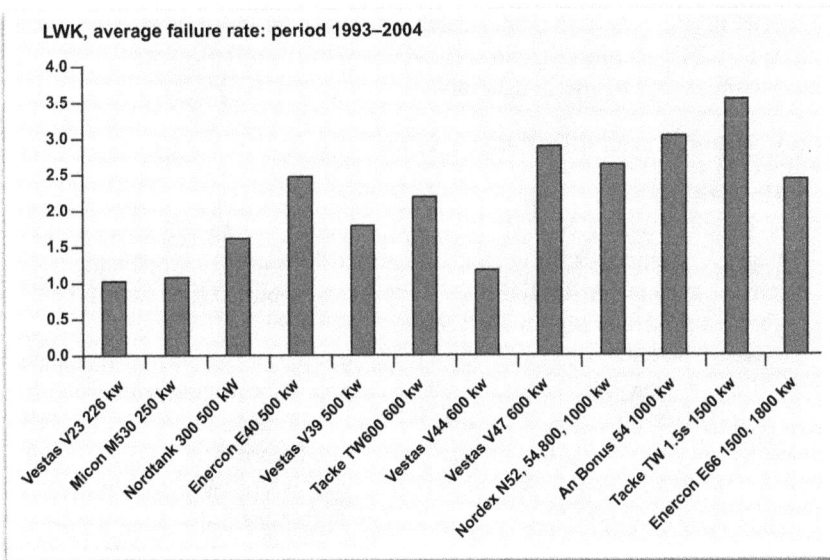

Figure 5.1 Reliability data on repairable on-shore WT failure rates against size, Tavner (2012)

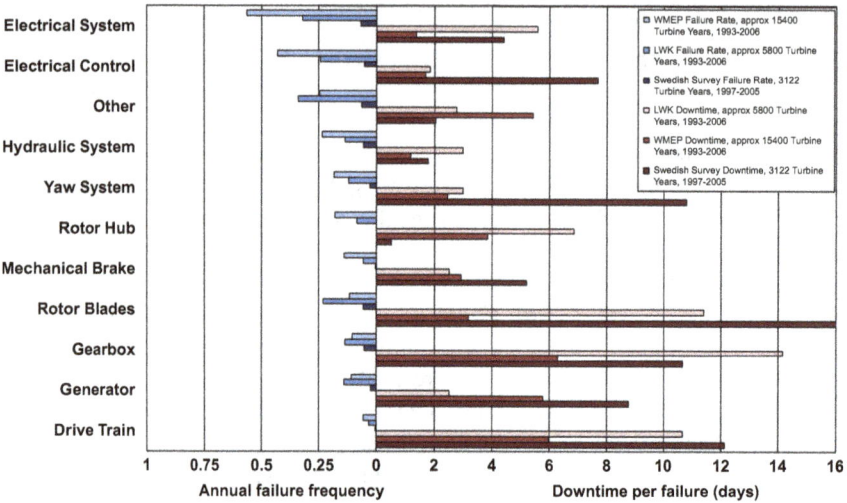

Figure 5.2   Reliability data on repairable on-shore WTs, failure rates and downtimes, Tavner (2012)

for subsea equipment failures are collected on a total life-time basis. The failure rates relate to generic sub-assemblies, which have physical boundary-defined parts, with detailed statistical measures of the sample population. The information collected is from equipment performing under normal operating conditions. The source data is stored in a computer database, access to which is only available to participating OREDA oil companies. However, some data is presented in the public domain in the form of generic data tables.

5.3.1.3   Generic reliability databases

- **MIL-HDBK-217F**

  The MIL-HDBK-217F (1991) was developed by the US Department of Defense. It comprises failure rate estimates for various components used in electronic systems. MIL-HDBK 217F provides empirical filed failure-rate data on both commercial and military electrical components, for use in reliability analyses. The collected data is presented from resistors, capacitors, inductors, transformers, integrated circuits and other largely electrical components.

  It has been observed that MIL-HDBK 217F failure rates, compared to OREDA (1984–2015), do not represent field failure data but are based on laboratory tests under controlled environmental stresses, e.g. temperature, humidity and voltage. MIL-HDBK 217F failure rates only relate to so-called primary component failures and do not account for external stresses or common cause failures. However, several tables with adjustment factors for data of MIL-HDBK-217F are established and published in RIAC (2010) to accommodate uncertainties.

- **NPRD-95; NPRD 2011**

  The NPRD-95 (1995) provides reliability and failure rate data on non-electronics: mechanical, electromechanical and discrete electronic parts and assemblies of 25,000 parts of military and commercial applications.

  The data has been updated as NPRD-2011 and gives a wider range of components (NPRD-2011). The NPRD-2011 disclose summary and detailed data sorted by part type, quality level, environment and data source. The data is compiled from field experience of military, commercial and industrial applications and focuses on component and sub-assemblies which have not been included in MIL-HDBK-217F (1991). The data is as follows: part descriptions, quality level, application environments, point estimates of failure rate, data sources, number of failures, total operating hours, miles, or cycles and detailed part characteristics (NPRD-2011). The MIL-HDBK-217F and the NPRD-95; 2011 are complementary to each other.

- **Integrated reliability database: IEEE gold book**

  An extensive survey in the United States, conducted by a group led by Dickson in the 1960s, was followed by several IEEE reliability surveys between 1973 and 1996, IEEE Gold Book (1999).The survey provides data on commercial power distribution systems. This survey included generators, power transformers, rectifier transformers, circuit brakes, disconnect switches, cables, cable joints and terminators, and electrical utility power supplies (IEEE Gold Book, 1997). The historical data provided can be used to compare alternative technologies.

5.3.1.4 Portfolio of surrogate data and useful references

From the above sources, the subsequent main sub-assembly sections of this chapter establish the portfolio of surrogate data (PSD) that will be used in this book for analysing the reliability of the above six devices.

The following references are particularly relevant:

- Carroll J, McDonald A, McMillan D (2014), Reliability comparison of wind turbines with DFIG and PMG drive trains;
- Chung, H S, Wang, H, Blaajberg, F, Pecht, M Eds (2015), Reliability of power electronic converter systems;
- Moir, I (1998), The all-electric aircraft-major challenges;
- OREDA (1984–2015), Off-shore & on-shore reliability data;
- Reliability Information Analysis Center, Reliability Modelling, RIAC (2010);
- Schmid, J, Klein, H P (1991), Performance of European wind turbines;
- Tavner, P J (2012), Off-shore wind turbines-reliability, availability & maintenance, including in particular:
  - ReliaWind data;
  - Windstats data;
- Tavner, P J, Faulstich, S, Hahn, B and van Bussel, G J W (2011), Reliability and availability of wind turbine electrical and electronic components.

5.3.2   Major sub-assemblies

5.3.2.1   Electrical machines

Information about the failure rates of electrical machines originally came from experience in the defence industry, where reliability predictions are a contractual requirement of equipment purchase, see Tavner *et al.* (2008). Information is also available from the wind industry about generators being fitted to the WTs, which are being installed in increasing numbers. Fault data for these generators are being recorded sufficiently frequently to deduce the life curves of typical electrical machines and their reliability.

The MTBF can be a deceptive quantity. It is intended to represent the prospective life of the device, assuming it has a constant failure rate, as shown in the constant failure rate region of Figure 3.5. This suggests a 50% probability of failure before the MTBF and 50% probability of failure afterwards. However, devices can have a failure rate that improves with time and then it is possible that a higher proportion of failures could occur before the MTBF. That said, a well-maintained device can expect to be operating in the constant failure rate region, in which case MTBF gives a good indication of prospective life, which engineers can appreciate without being overwhelmed by any statistical interpretation.

Table 5.1 extracts data from a number of electrical machine surveys giving the failure rates and MTBFs for a range of electrical devices, facilitating their suppression. There is a remarkable degree of consistency, with MTBFs ranging from 18 to 33 years, if a year is assumed to contain 8,766 hr operation. The table also gives an idea of the significance of each survey, by noting the number of devices surveyed and the number of failures recorded. It should be noted that the large surveys are dominated by induction motors, because of their ubiquity.

The distribution of failures within the configuration of the machine is also important because it guides the direction in which condition monitoring is applied. Table 5.2 gives an analysis of failures, based on the literature, and three important areas of the machine are identified:

• Stator related, primarily the winding;
• Rotor related, including slip rings and bearings;
• The remainder of failures are grouped as other.

The data comes from different surveys, and whilst these surveys are not necessarily complementary, they are substantial and do show consistent failure areas, which are, in descending order of importance:

• Bearings;
• Stator related;
• Rotor related.

The relative importance of failures is affected by the size, voltage and type of machine under consideration. In particular, the relative weighting of stator to rotor winding failures depends upon the type and size. For example, small, low voltage induction machine failures, exemplified by the first two columns of Table 5.2, are

Table 5.1 *Typical measured electrical machine failure rates and MTBFs,*
obtained from literature

Type of machine	Number of machines in survey	Machine years in survey	Number of failures in survey	Failure rate, failures/ machine/ yr	MTBF, hrs	Reliability survivor function, $R_{(1yr)}$, %	Source of data
Large steam turbine generators	Not known	762	24	0.0315	278,130	97	IEEE, Gold Book (1997)
Induction motors 601–15,000 V	Not known	4,229	171	0.0404	216,644	96	
Motors greater than 200 hp, generally MV and HV induction motors	1,141	5,085	360	0.0708	123,735	93	
Motors greater than 100 hp, generally MV and HV induction motors	6,312	41,614	1,474	0.0354	247,313	97	
Motors greater than 11 kW generally MV and HV induction motors	2,596	25,622	1,637	0.0639	137,110	94	
WT generators < 2 MW	Various	44,785	4,389	0.0980	89,386	91	Spinato et al. (2009)

dominated by bearings, as low voltage windings experience very few faults. Smaller electrical machines use greased or oil-lubricated rolling element bearings and their reliability depends heavily on the standard of maintenance. Larger electrical machines use oil-lubricated sleeve bearings, which have a much higher intrinsic reliability. Induction machines show a much lower number of rotor winding or squirrel cage faults, compared to stator winding faults, because of the ruggedness of the cage construction. But larger, high voltage machines, exemplified by the next three columns of Table 5.2, receive a higher proportion of failure modes on the stator winding, due to dielectric stress and vibration root causes. This can be of a similar significance to bearing faults. Large machine bearings are also usually of sleeve construction and, with constant lubrication, are generally more reliable than bearings in smaller machines. Table 5.2 also shows that other

Table 5.2   Electrical machine distribution of failed sub-assemblies, obtained from literature

Sub-assemblies	Predicted by an OEM through FMEA techniques*	MOD survey, Tavner *et al.* (2008)	IEEE large motor survey, IEEE Gold Book (1999)	Motors in utility applications, IEEE Gold Book (1999)	Motor survey off-shore and Petrochem, IEEE Gold Book (1999)
Types of machines	Small to medium LV motors and generators <150 kW, generally squirrel cage induction motors	Small LV motors and generators <750 kW, generally squirrel cage induction motors	Motors greater than 200 hp generally MV & HV induction motors	Motors greater than 100 hp generally MV and HV induction motors	Motors greater than 11 kW generally MV and HV induction motors
Bearings	75%	95%	41%	41%	42%
Stator related	9%	2%	37%	36%	13%
Rotor related	6%	1%	10%	9%	8%
Other	10%	2%	12%	14%	38%

*Private communication from Laurence, Scott & Electromotors Ltd.

component failure modes, for example in the cooling system, connections and terminal boxes, become more significant on larger machines.

Induction machine cage faults can be detected through perturbations of the air gap magnetic field. Considerable scientific work has gone into that detection, probably due to the fact that the air gap field is scientifically interesting, analytically tractable and cage induction motors are very numerous. But these faults are not as significant as the scientific community seems to think.

Bearing faults, on the other hand, involve a more complex interplay of mechanical and electrical physical effects, leading to air-gap eccentricity, which affects the resultant magnetic field. This is also tractable, although more complex than the effect of rotor cage faults. Some of the lessons learnt from rotor cage fault detection, combined with the study of the effect of eccentricity on the air gap magnetic field, can be applied to the detection of bearing faults. It is interesting that the study of the effects of eccentricity in induction motors also addresses issues of machine noise and speed control.

5.3.2.2 Gearboxes

There is surprisingly little data available on gearbox reliability, which must depend on the rating, cooling type, number of stages of the gearbox and the frequency with which failure data is collected. However, the data given in Table 5.3 has been obtained from the European wind industry, Tavner (2012) and Spinato *et al.* (2009).

5.3.2.3   Power converters

More information is available on converter failures, Table 5.4, because of the wide range of safety-related industries using converter technology. The data in Table 5.4 is taken from the extensive European wind and aerospace industries.

Table 5.3   Summary of gearbox failure rates

Type of device	Failure rate, failures/machine/yr	MTBF, hr	Reliability survivor function, $R_{(1yr)}$, %	Source
Gearbox	0.1550	56,516	86	Tavner (2012)
	0.2000	43,800	82	Spinato *et al.* (2009)
	0.3000	29,200	74	

Table 5.4   Summary of converter failure rates

Type of device	Failure rate, failures/machine/yr	MTBF, hr	Reliability survivor function, $R_{(1yr)}$, %	Source of data
VSC WT converter	0.045	194,667	96%	Spinato *et al.* (2009)
	0.243	36,049	78%	
Soft starter with six thyristors and power factor correction capacitors, assembly failure rate estimate based on surrogate WT data	0.063	139,048	94%	Chung *et al.* Eds (2015)
LV partially rated converter (PRC) with six IGBT/diodes per inverter, assembly failure rate estimate based on surrogate WT data	0.121	72,397	89%	
Three-axis pitch converters with two IGBT/diode choppers per converter, assembly failure rate estimate based on surrogate WT data	0.195	44,923	82%	
MV fully rated converter (FRC) with 12 IGCT/diodes per inverter in two parallel PRC, assembly failure rate estimate based on surrogate WT data	0.402	21,791	67%	
MV fully rated converter (FRC) with 12 IGCT/diodes per inverter in one single channel, assembly failure rate estimate based on surrogate WT data	0.738	11,870	48%	
LV partially rated VSC (PRC) in the power range 0.6–0.8 MVA to convert generator outputs to fixed LV power frequencies	0.100	87,600	90%	Carroll, *et al.* (2014)
LV fully rated VSC (FRC) in the power range 2–3 MVA to convert generator outputs to fixed LV power frequencies	0.593	14,772	55%	
VSC aircraft converter	2.190	4,000	11%	Moir (1988)
VF aircraft converter	0.438	20,000	65%	
VSCF aircraft cyclo-converter	0.701	12,500	50%	

5.3.2.4   Electrical, cable and connector systems

More information is available on electrical failures, Table 5.5, because of the substantial reliability work of the IEEE taken from the very extensive power generation and electrical transmission and distribution industries, primarily in the United States.

Table 5.5   Summary of electrical failure rates

Type of device	Failure rate, failures/ machine/ yr	MTBF, hr	Reliability survivor function, $R_{(1yr)}$, %	Source
LV contactor	0.0052	1,684,615	99	IEEE
LV circuit breaker feeding auxiliary supply	0.0027	3,244,444	100	Gold Book
LV circuit breaker feeding LV export cable	0.0044	1,990,909	100	(1999)
Battery electrical auxiliary system for MEC	0.0319	274,725	97	Delorm (2014)
Flexible LV transmission cable, assume 500 m length	0.0012	7,423,729	100	IEEE Gold Book (1999)
Buoyant LV or MV cable for MEC application, assume 100 m length	0.0422	207,614	96	Delorm (2014)
Cable buoyancy support with swivel	0.0000014	6,060,606,061	100	
Sub-sea connector	0.0236	371,186	98	
Fixed LV export cable systems, assume 10 km length	0.0236	371,186	98	IEEE Gold
Dry type LV/MV transformers to raise export cable voltage for power output to shore-side sub-station	0.0036	2,433,333	100	Book (1999)
MV or LV isolator	0.0052	1,684,615	99	
MV contactor	0.0153	572,549	98	
MV circuit breaker feeding the sea-shore export cable or shore-side sub-station or MEC output to power grid	0.0064	1,368,750	99	
Fixed MV export cable systems, assume 10 km length	0.0052	1,684,615	99	
Dry type MV/LV transformer to step down output voltage for LV auxiliary system	0.0018	4,866,667	100	
Oil-cooled MV/HV transformers to raise export cable voltage for power output to the grid	0.0030	2,920,000	100	

5.3.2.5 Hydraulics and ancillary systems

Some information that is available on hydraulic system failures is not extensive and is dominated by the European wind industry, Table 5.6, so this data needs to be treated with caution.

5.3.2.6 Structures, anchors and ancillary systems

Information on the reliability of structural and anchor systems for MECs, Table 5.7, is difficult to locate, and there is an argument, see Section 3.6.2 and Sorensen, J in

Table 5.6   Summary of hydraulic failure rates

Type of device	Failure rate, failures/ machine/yr	MTBF, hr	Reliability survivor function, $R_{(1yr)}$, %	Source
Hydraulically operated brake system	0.0550	159,273	95	Delorm (2014)
Ventilation system	0.1048	83,612	90	Delorm (2014)
Water-cooled heat exchanger	0.0936	93,633	91	Delorm (2014)
Hydraulic power units for MEC application	0.0394	222,222	96	Delorm (2014)
Hydraulic receivers, accumulators and tanks for MEC application	0.0394	222,222	96	Delorm (2014)
Sub-sea hydraulic rams	0.0090	973,333	99	Tavner (2016)
Yoke for hydraulic rams	0.0045	1,946,667	100	Tavner (2016)

Table 5.7   Summary of structural failure rates

Type of device	Failure rate, failures/ machine/ yr	MTBF, hr	Reliability survivor function, $R_{(1yr)}$, %	Source
Civil engineering steel or concrete structures and foundations for MEC application	0.0011	7,963,636	100	Delorm (2014)
Monopile or fixed sea-bed location and pinning systems for MEC application	0.0011	7,963,636	100	
Nacelle or housing or duct structure	0.0011	7,963,636	100	
Cross-beam structure	0.0011	7,963,636	100	
Detachable sea-bed location and pinning systems for MEC application	0.0022	3,981,818	100	Tavner (2016)
Buoyant, hinged steel structures for MEC application	0.0022	3,981,818	100	
Buoyant hull for MEC application	0.0011	7,963,636	100	Delorm (2014)
Retractable turbine legs	0.0090	973,333	99	

(Continues)

Table 5.7   (Continued)

Type of device	Failure rate, failures/ machine/ yr	MTBF, hr	Reliability survivor function, $R_{(1yr)}$, %	Source
Mooring yoke system	0.0500	175,200	95	
Connecting links	0.1000	87,600	90	
Catenary mooring, chain	0.0237	370,022	98	
Catenary mooring, steel wire rope	0.0000	284,323,320	100	
Pile anchors	0.0074	1,186,250	99	
Corrosion protection	0.0443	197,686	96%	

Steenbergen *et al.* (2014), that a more sophisticated analysis would be appropriate. However, at the present time, the methodology of Chapter 4 has been applied and the data is dominated by the European wind industry, Table 5.7, so this data needs to be treated with caution.

5.4   Summary

This chapter summarises the knowledge available on the failure rates of engineering sub-assemblies used in MECs. It demonstrates that there is a great deal of reliability information already available, some of it from a renewable energy environment similar to that into which MECs will be deployed.

However, there are important gaps in that knowledge, notably in the areas of hydraulics, structural components, sub-sea components and sea-shore cabling systems.

These gaps are likely to be filled by the wide range of research which is going on at present and by the evidence acquired from operational MEC demonstrators, even if a great deal of that information is anecdotal until industry standards are developed.

Chapter 6

Reliability and MEC device taxonomy

6.1 Introduction

The purpose of this chapter is to describe the taxonomy of several typical wave and tidal MEC configurations and make comparison with recent wind turbine installations.

The MEC configurations have been chosen to exemplify the most important reliability principles; however, none of these examples have yet continued into long-term operation. That is because many lessons still need to be learnt in the marine environment. Many lessons are now being applied from this experience, arising from understanding the impact of key sub-assemblies on reliability. Also from considering how their sub-assemblies could best be arranged to improve the reliability necessary from off-shore wave and tidal devices.

The author's intent has been to develop a standard way of presenting various devices, so that the readers can contrast and compare results by applying this methodology to future prototypes under consideration.

These models could then be used to apply well-known reliability modelling techniques, described in Chapter 3, to improve these and future models.

This type of analytical approach has been used to progressively improve the reliability of many industrial products, for example:

- Electricity production and transmission;
- Railway locomotives and train-sets;
- Aircraft propulsion;
- Wind turbines.

So the six MEC devices to be considered are as follows:

- Wave energy converters:
  - An on-shore oscillating water column device, based loosely upon the Wavegen, Limpet concept,
    WECa;
  - A near-shore oscillating wave surge converter, based loosely upon the Voith, Aquamarine, Oyster concept,
    WECb;
  - An off-shore surface-following attenuator, based loosely upon the Pelamis P2 concept WECc.

- Tidal stream devices:
  - A fixed mono-pile device, incorporating two horizontal-axis, pitched-blade, geared turbines, based loosely upon the Siemens SeaGen, marine current turbine concept,
    TSDa;
  - A sea-bed device, incorporating a ducted, horizontal-axis, fixed-pitch, direct-drive turbine, based loosely upon the DCNS, open hydro concept,
    TSDb;
  - A moored, two horizontal-axis, fixed-pitch, geared turbine, based upon the Scotrenewables concept,
    TSDc.

Figure 6.1 shows these six MECs in diagrammatic form;

- Figure 6.1(a) illustrates the wave energy converters;
- Figure 6.1(b) illustrates the tidal system devices.

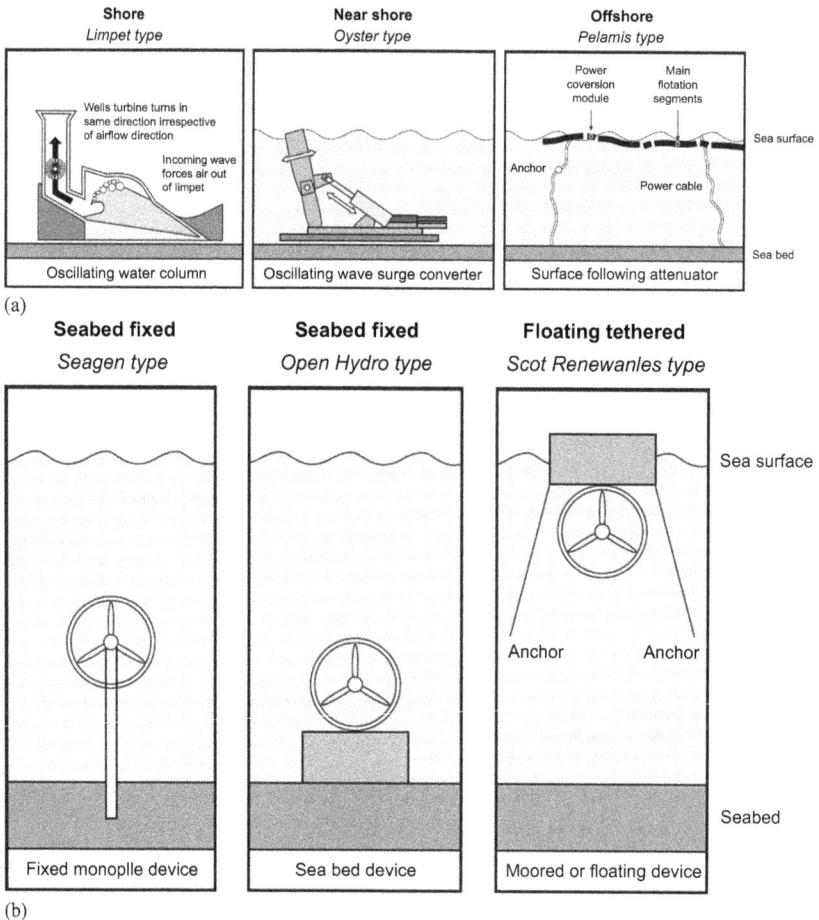

Figure 6.1   Diagrammatic representations of the MECs to be considered: (a) wave energy converters, WECa, WECb and WECc and (b) tidal system devices, TSDa, TSDb and TSDc

The following sections show examples of these devices, their characteristics and design features. The following chapters will dissect these features, to facilitate more penetrating reliability analysis for the future development of renewable energies.

This is not a reliability analysis of specific devices, since the analysis is based upon generic information only, and not upon detailed device design information, which is the confidential intellectual property of the companies concerned. Nor is this device selection a beauty contest of likely or preferred candidates, or an indication of their future development. Rather, it is the selection of six existing devices that have extracted energy from the marine environment and that each adopt diverse configurations, which can be evaluated to determine general principles about design for reliability. They represent some of the most well-developed marine energy devices, with most of the key electrical, mechanical and structural features that must be incorporated into future successful MECs. In addition, sufficient information is available to allow early reliability modelling to be undertaken.

The intention is that future MEC device designers will use these approaches to evaluate specific details of their own products, make reliability comparisons with their competitors and produce successful devices.

6.2 Wave energy converter (WEC) device configurations

6.2.1 Limpet

See Figure 6.2 and Table 6.1 for Limpet concept.

6.2.2 Oyster

See Figure 6.3 and Table 6.2 for Oyster concept.

6.2.3 Pelamis

See Figure 6.4 and Table 6.3 for Pelamis concept.

Figure 6.2   Limpet, Prototype 150 kW WEC demonstrator producing power in Islay 2003

Table 6.1   Characteristics and design features of WECa loosely based on Wavegen, Limpet

Features	Design characteristics
Reinforced concrete structure on fore-shore: • Sea-bed location and pinning system Turbine power house and sub-station containing: • Wells turbine(s); • Drive shaft(s) and couplings; • Direct drive induction generator(s); • LV VSC(s); • Switchgear to parallel VSCs if required; • Transformer to raise VSC output from LV to MV for power output to grid; • VSC controller to control WEC output; • SCADA monitoring interface; • Circuit breaker feeding grid	A fixed, hollow, open, on-shore concrete structure whose aperture is closed by the waves Wave pressure moves columns of air through one or more bi-directional fixed blade wells turbines, each coupled to a synchronous or asynchronous generator mounted in the device power house As the air column rises and falls, it pressurises and de-pressurises the turbine(s), which spin(s) in one direction and generate variable frequency LV electricity in the coupled generator(s) Generator(s) are coupled by LV VSCs via circuit breakers to the fixed frequency electrical grid via a shore sub-station, where voltage is stepped up from LV to MV and the output can be paralleled with other WECs Power flow to the grid is regulated by the overall WEC control strategy and the sub-station controller

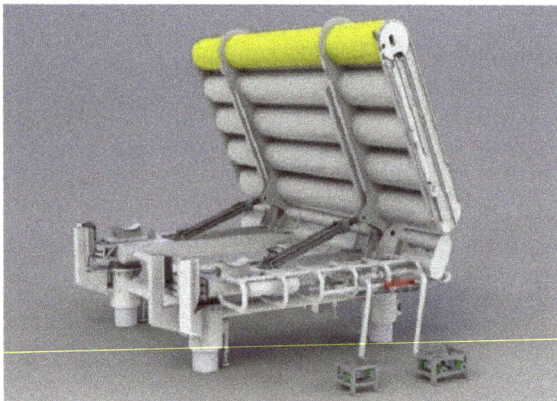

Figure 6.3   Oyster, Prototype 800 kW full-scale WEC demonstrator, began producing power at EMEC Orkney, UK, 2009

Table 6.2   Characteristics and design features of WECb, loosely based on Oyster

Features	Design characteristics
Buoyant flap structure incorporating: • Buoyant flap and hydraulic rams; • Sea-bed base and pinning; • Base accumulators and infrastructure for WEC to shore pipelines; • Fresh-water pipeline connections; • Buoyant flap mounting and detachment device; • SCADA monitoring interface	A buoyant flap, weighted and pinned to the sea-bed operates one or more near horizontal hydraulic rams As each wave passes, each ram pressurises a fresh water accumulator system to pump water through device-to-shore pipework
Shore-side power house and sub-station containing: • Fresh-water pipeline connections; • Fresh-water turbine(s); • Direct drive LV induction or synchronous generator(s); • LV VSC(s); • Switchgear to parallel VSCs if required; • Transformer to raise VSC output from LV to MV for power output to grid; • VSC controller to control WEC gain and output; • SCADA monitoring interface; • Circuit breaker feeding grid	Low-speed turbines housed in an on-shore power house operate in the pressurised water pipework. Each turbine is coupled to a synchronous or asynchronous generator to generate LV electricity Generator(s) are coupled by LV VSCs via circuit breakers to the fixed frequency electrical grid, via a shore sub-station, where voltage is stepped up from LV to MV and the output can be paralleled with other WECs Power flow to the grid is regulated by the overall WEC control strategies and the sub-station controller

Figure 6.4   Pelamis P2, 750 kW WEC in grid-connected operation EMEC Orkney, UK, 2014

Table 6.3   Characteristics and design features of WECc, loosely based on Pelamis P2

Characteristics	Design features

Four articulated buoyant hulls each incorporating:
• Bow mooring and detachment device;
• Stern buoyant export cable and detachment device;
• Sway and heave hinges between the buoyant hull and power take-off units;
• Three power take-off units each incorporating:
  • Four hydraulic rams for roll, pitch and yaw;
  • Hydraulic accumulators, hydraulic receivers and tanks;
  • Two hydraulic motors;
• Two LV induction generators;
• LV voltage source converters (VSC);
• Controller for WEC gain and output;
• SCADA monitoring interface;
• Circuit breaker feeding export cable
	Four buoyant hulls are connected by three hinged, flexible power take-off units via heave and sway hinges
	The assembly of hulls and power take-off units is moored to the sea-bed at the bow and electrically connected to the shore by a buoyant suspended export cable near the stern
	Each power take-off unit incorporates flexible hydraulic ram couplings to control the vertical and lateral stiffness of the device
	As the hinged buoyant hulls move through the sea-way of an oncoming wave field, the ram couplings pressurise a hydraulic system and accumulators in the adjacent hull, circulating hydraulic fluid through two separate hydraulic motors, driving induction generators
	Each hydraulic motor/generator pair in each hull is electrically paralleled to the motor/generator pairs in neighbouring hulls

Shore-side sub-station containing: • Transformer to raise VSC output from LV to MV for power output to grid; • VSC controller to control WEC output; • SCADA monitoring interface; • Circuit breaker feeding grid	LV electrical systems of each hull are connected to the export cable via a circuit breaker from the device to the shore-side sub-station, where voltage is stepped up from LV to MV and the output can be paralleled with other WECs Power flow from each power take-off unit is regulated by the WEC device controller to manage accumulator pressures, device gain and redundancy Power flow to the grid is regulated by the overall WEC control strategies and the sub-station controller

6.3   Tidal stream device (TSD) configurations

6.3.1   Sea-Gen

See Figure 6.5 and Table 6.4 for Sea-Gen concept.

6.3.2   Open hydro

See Figure 6.6 and Table 6.5 for Open Hydro concept.

Figure 6.5   Sea Gen, Prototype 1 MW TSD in the Water at Strangford Loch, UK, 2008

Table 6.4   Characteristics and design features of TSDa, loosely based on SeaGen

Features	Design characteristics
Monopile tower, piled into the sea-bed incorporating: • Monopile; • Moveable beam supporting two turbine nacelles; • Hydraulic power unit to raise and lower the turbine beam; • VSC to convert generator output to fixed frequency LV power; • Transformer to raise VSC output from LV to MV for power output; • VSC Controller to control TSD output; • SCADA monitoring interface to shore; • Circuit breaker feeding export cable; • J tube for export cable; • Export cable.	A fixed monopile tower, piled into the sea bottom, carries a horizontal, vertically movable beam mounting two matching, horizontal axis, variable pitch turbines The moveable beam can be raised or lowered hydraulically into the tideway, using a tower-mounted hydraulic power packThis can raise the turbines above the sea surface for maintenance purposes Each turbine drives an LV induction generator through a 2-stage gearbox mounted in the turbine hub or nacelle, which feeds electricity from the turbines to the tower
Two hydraulic variable pitch turbine nacelles incorporating: • Two 2-blade, variable speed, pitch-controlled turbines;	In the turbine tower the two electrical supplies are converted to mains frequency by parallel LV VSCs, and voltage is stepped up from LV to MV in a transformer The transformer feeds MV electrical power out

(Continues)

Table 6.4   (Continued)

Features	Design characteristics
• Two hydraulic pitch control units; • Two water-cooled 2-stage gearboxes; • Two water-cooled LV induction generators.	from the monopole, via a J tube, an export cable and circuit breakers to a shore-side substation, where the output can be paralleled with other TSDs and stepped-up from MV to HV
Shore-side sub-station containing: • Transformer to raise export cable outlet from LV to MV for power output to grid; • SCADA monitoring interface; • Circuit breaker feeding grid.	The power flow to the grid is regulated by the control system of the converters and the overall control strategy of the TSD

Figure 6.6   Open Hydro, Prototype 250 kW TSD raised above the water before grid-connected operation at EMEC Orkney, UK, 2006

6.3.3   Scot Renewables

See Figure 6.7 and Table 6.6 for Scot Renewables concept.

6.4   Device configuration taxonomy

6.4.1   General concepts and configurations

The previous section introduced six devices, three WECs and three TSDs, that incorporate the main features of all MECs. The following sections will show the

Table 6.5   Characteristics and design features of TSDb, loosely based on Open Hydro

Features	Design characteristics
Fixed structure pinned to sea-bed, incorporating: • Sea-bed location and pinning system; • Housing for the turbine; • Internal cabling; • Export cables to shore.	A structure mounted fixed to the sea bottom carrying a ducted, horizontal axis, fixed-pitch sea-water-lubricated turbine, with sealed permanent magnet rotor, rotating in a sea-water with a sealed, wound-stator as rim generator.
Single fixed-pitch, variable-speed turbine incorporating: • Turbine blades and rim permanent magnet rotor; • Sea-water hydraulic bearing between stator and rotor; • Water-cooled rim generator stator; • Internal stator cabling.	In the example shown the turbine can be raised or lowered hydraulically into the tideway, using tower-mounted hydraulic power packs. This raises the turbine above the sea surface for maintenance purposes. However, the future design intent is that larger devices would be sea-bed mounted with no raising device.
Shore-side power house • VSC to convert generator output to fixed frequency LV power; • Off-shore device controller; • Transformer to raise VSC output from LV to MV for power output; • VSC controller to control TSD output; • SCADA monitoring interface to shore; • Circuit breaker feeding from export cable.	The turbine directly drives an LV permanent magnet generator, which feeds variable frequency LV electricity from the turbine to the shore via an LV export cable. Electrical supply from the export cable is converted to mains frequency by LV VSC in the shore-side power house, where other TSD devices can be paralleled and voltage stepped up from LV to MV in a transformer in the shore-side power house. The power flow to the grid is regulated by overall control strategy of the TSD converters in the shore-side power house.

Figure 6.7   Scot Renewables, Prototype 250 kW TSD preparing for deployment at EMEC Orkney, UK, 2010

Table 6.6   Characteristics and design features of TSDc, loosely based on Scot Renewables

Features	Design characteristics
Buoyant hull, incorporating: • Hydraulic power unit to deploy two turbine nacelles; • Two retractable turbine operating mechanisms; • VSC to convert generator outputs to fixed frequency LV power; • Transformer to raise VSC output from LV to MV power output; • VSC controller to control TSD output; • SCADA monitoring interface sea-shore; • Cabling; • Circuit breaker feeding export cables to shore; • Bow mooring and detachment device; • Buoyant stern export cable and detachment device.	The device has a single buoyant hull, carrying two turbine units which can be retracted for towing to the off-shore tidal site and then be deployed when in position in the tide-way. The hull is moored to the sea-bed at the bow and electrically connected to the shore by a buoyant suspended export cable near the stern. As the buoyant hull sits in tideway the oncoming tidal field operates the turbines and the device. When the tide changes from ebb to flood, the device passively veers or backs around its mooring point to generate in the opposite tidal stream. Each variable-speed, fixed-pitch turbine drives a permanent magnet synchronous generator, through a two-stage gearbox, mounted in the turbine hub or nacelle.
Two hydraulic, retractable, variable-speed, turbine nacelles incorporating: • Two pitch-controlled, variable-speed, • Two-blade turbines; • Two hydraulic pitch-control units; • Two water-cooled 2-stage gearboxes; • Two water-cooled LV induction generators.	Each generator feeds LV power from each nacelle to the hull, where the two electrical supplies are paralleled and converted to mains frequency by LV VSCs. The voltage is stepped up from LV to MV in a transformer. The transformer feeds MV electrical power out from the hull, via the export cable and circuit breakers, to a shore-side substation. Here the output can be paralleled with other TSDs and stepped-up from MV to HV.
Shore-side sub-station incorporating: • Transformer to raise export cable output from LV or MV to HV for power output to grid; • Sea-shore SCADA monitoring interface; • Circuit breaker feeding from export cable.	The power flow from the device export cable is regulated by the converter control systems and the overall control strategy of the TSD. The power flow to the grid is controlled by the sub-station control system.

variance in engineering of these concepts. This will serve as a preliminary to studying their failure rates and then making predictions of their reliability.

A wide range of concepts and configurations have been introduced, and a cursory view of Tables 6.2 and 6.3 show that there are many more being considered.

However, by concentrating on the 6 considered here, the main elements of MEC configuration can be identified.

6.4.2 Systems, populations and operating experience

Section 6.2 has shown that there are key systems that appear in almost all MECs. Many of them are in widespread use in the conventional and renewable power generation and marine environments, although some are being used in different formats for off-shore wind. They are all described below, the most novel aspects being indicated by *:

- *Civil engineering steel or concrete structures for MEC application;
- *'Monopile or fixed sea-bed location and pinning systems for MEC application;
- *Detachable sea-bed location and pinning systems for MEC application;
- *Buoyant, hinged steel structures for MEC application;
- *Buoyant structure mooring and detachment devices for MEC application;
- *Turbines for MEC application;
- *LV export cables and detachment devices for buoyant cables for MEC application;
- Fixed LV or MV export cable systems;
- *Hydraulic power units for MEC application;
- *Hydraulic receivers, accumulators and tanks for MEC application;
- Hydraulic pitch control units;
- *Variable or fixed speed, pitch-controlled or uncontrolled turbines for MEC application;
- Water-cooled or air-cooled gearboxes;
- Water-cooled induction or wound or permanent magnet synchronous generators;
- VSCs to convert generator outputs to fixed LV power frequencies;
- VSC controllers;
- *Device controllers and output gain systems for MEC application;
- *SCADA monitoring interfaces, sea-shore or shore-sea, for MEC application;
- Shore-side sub-stations;
- LV Circuit breakers feeding the export cable sea-shore or the shore-side power grid;
- LV/MV or MV/HV Transformers to raise export cable voltage for power output to the grid;
- MV or HV Circuit breakers to feed the MEC output to the power grid;
- Sub-station power flow control systems.

6.5 Analysis of device concepts

Delorm (2014) presented a detailed method for assessing the reliability of TSDs, which will be applied to the six selected WECs and TSDs described above.

It may be noted that the three TSDs were included in analyses previously reported by Delorm *et al.* (2011) and Delorm (2014).

The results presented here differ slightly because more data has become available and small modifications to the models, based on greater knowledge of device architectures, have come to light.

However, that does not invalidate the Delorm (2014) proposition:

- For each device, reliability models can be obtained by developing a schematic diagram;
- From which a functional and reliability block diagram (FBD and RBD) can be prepared;
- Hence a table of reliabilities for all assemblies, sub-assemblies and components can be compiled.

This process is rather like building a Lego castle for each device.

There are many pieces, like Lego blocks of different colours, that can be put together, with different reliability data. Every time you construct such a model there are small but relatively insignificant differences from a reliability point of view.

The important issue is that the main features are captured and that differences between devices can be identified, so in the results their impact on reliability behaviour is reflected. By applying this approach systematically across all six devices, using a common list of component reliabilities or failure rates, it is possible to make a comparison between the likely prospective reliabilities of whole devices and thereby of those concepts.

The results of these analyses, using the above methodology, are presented in Appendix B with their FBDs, RBDs and Spreadsheet calculations of reliability.

The purpose of these analyses is to draw attention both to the reliability similarities between WECs and TSDs and their differences between, which will be elaborated in the following chapters.

Table 6.7 and Figures 6.8–6.10 give a summary of the results of this reliability analysis.

To provide a basis for comparison, data has also been assembled using a similar method to off-shore WTs, taken from Delorm *et al.* (2016).

This may be compared with WECs, TSDs and on-shore WTs, also as reported in Delorm (2013).

So the results are summarised in Table 6.7 and shown graphically in Figures 6.8–6.10, where on-shore WTs failure rates were 2.5–3.5 failures/yr.

Figure 6.8 shows the number of sub-assemblies in each device.

The off-shore WT reliability analysis was done with only 18 or 19 sub-assemblies considered for each device and no redundancy effects being considered.

The analysis of the TSDs and WECs was done with between 30 and 90 sub-assemblies.

The following conclusions can be drawn from the Lego reliability block approach in Figures 6.8–6.10:

- The predicted reliability for large off-shore WTs, based on the methodology in this book, is similar to but lower than measured on-shore WT values.

Table 6.7   Summary of results based on methods described in this chapter

Devices	Device label	Based on	Sub-assembly numbers used in prediction	Sources	Data provided
Wave energy converters	WECa	Limpet, 150 kW	37	Various	Estimated λ_{max} and λ_{min} failure rates using methodology of Chapter 6, including the effects of redundancy
	WECb	Oyster, 800 kW	61		
	WECc	Pelamis, 750 kW	88		
Tidal stream devices	TSDa	Sea Gen, 1 MW	47		
	TSDb	Open Hydro, 250 kW	33		
	TSDc	Scot Renewables, 250 kW	64		
Off-shore wind turbines	OWTa	Siemens, SWT 3.6, 3.6 MW	17	Delorm et al. (2016)	
	OWTb	Vestas V90, 3 MW	18		
	OWTc	New Product, 6 MW	19		
On-shore wind turbines	WT	Various products		Tavner (2012)	λ measured in service

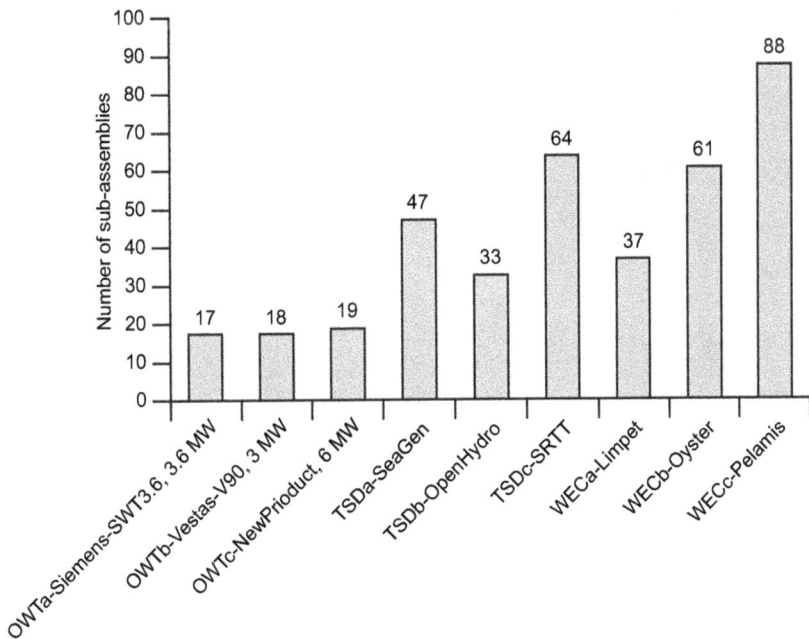

Figure 6.8   Numbers of sub-assemblies in off-shore WT, TSD and WEC devices

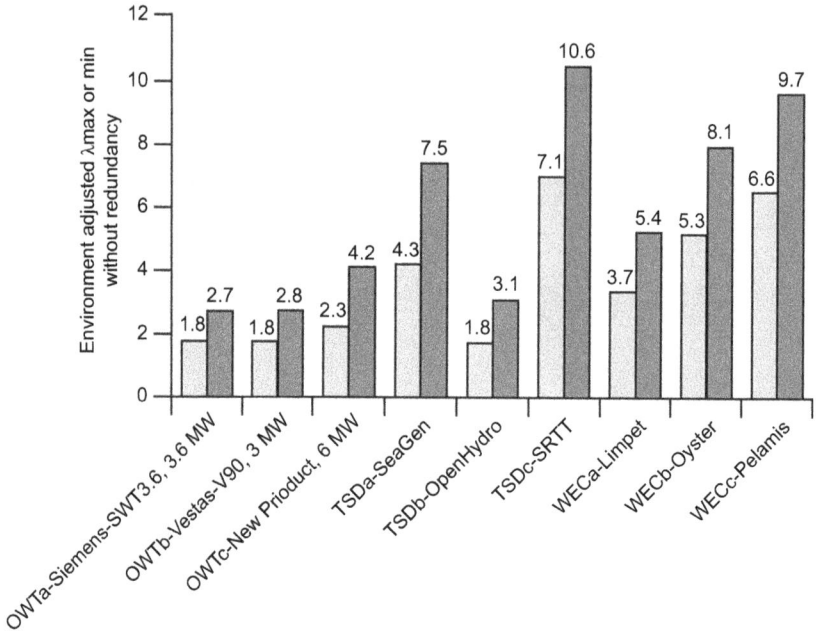

Figure 6.9   Summary of estimated λ_{max} and λ_{min} failure rates for off-shore WTs, TSDs and WECs, without redundancy

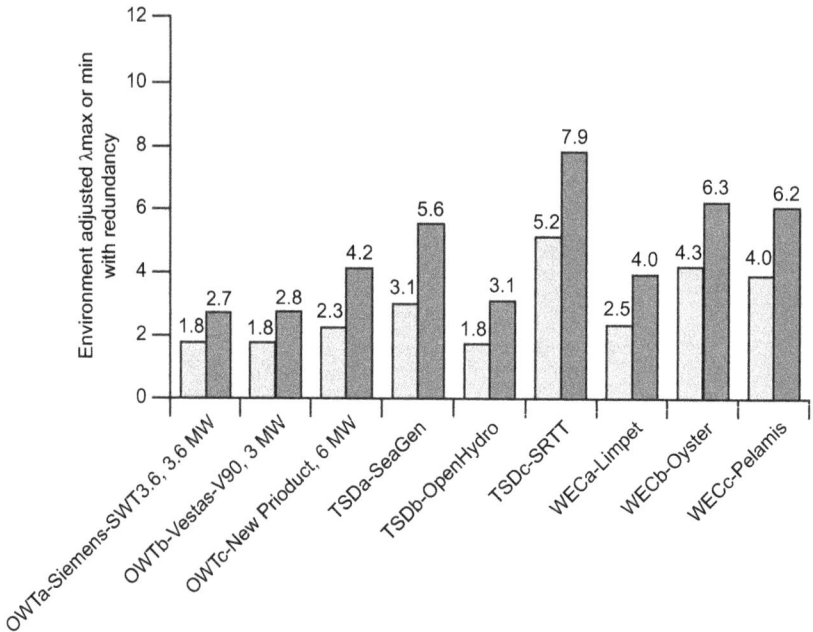

Figure 6.10   Summary of estimated λ_{max} and λ_{min} failure rates for off-shore WTs, TSDs and WECs, with redundancy

However, the number of sub-assemblies used to obtain those predictions was rather low;

- In general, the larger the number of sub-assemblies, the lower the reliability prediction, e.g. TSDc;
- In general, the selected TSDs have lower predicted reliabilities than off-shore WTs but similar predicted reliabilities to the selected WECs;
- Devices with fewer sub-assemblies have higher reliabilities, e.g. TSDb and WECa;
- Note that TSDb had no options to improve reliability with redundancy;
- Note how redundancy improves the reliability of those devices where it is an option and devices with higher levels of redundancy, e.g. TSDa and WECc, tend to have better predicted reliabilities, even though they may have a high sub-assembly count;
- Interestingly TSDc has a high level of redundancy but its reliability prediction is relatively poor because of a high sub-assembly count;
- Floating and moored detachable MECs, TSDc and WECc tend to have larger numbers of sub-assemblies and therefore a lower predicted reliability, even though they may prove easier to repair because they are detachable;
- Devices that concentrate their complex sub-assemblies on-shore, e.g. TSDb and WECa, tend to have a higher reliability.

6.6 Summary

Probabilistic reliability modelling is a proven method for assessing the probability of system success or failure at the conceptual design stage (Mackie 2008, RIAC&DACS 2005).

This chapter has presented the basis for converting generic MEC concepts into reliability prediction models to quantify MEC reliabilities in order to provide life-cycle predictions for specific types based on the method of Delorm (2014).

Surrogate reliability data from WT, marine and conventional electric power generation industries has been used to predict device failure rates and survivor functions. The work includes a systematic review of literature and data available on similar technology reliability, prior to carrying out a meta-analysis of these data sources.

There has been some criticism of reliability analysis, particularly from WTs, on the basis that the quantity of data available in the public domain may not be statistically significant; this is certainly a concern. However, in the absence of large volumes of operator reliability data, due to intellectual property limitations, the author has used the best data and methods currently available, with the understanding that, when OEMs and operators release sufficient data to meet statistical significance criteria, the results presented here may need to be reviewed.

The purpose of this chapter has been to provide a means of comparing relative concept reliabilities, rather than to derive definitive device failure rates. From the analyses presented, the following conclusions can be drawn:

- The predictions of reliability for OWTs are similar to those measured for on-shore WTs;
- TSDs are showing reliabilities worse than OWTs, but better than WECs;
- Devices with a higher degree of redundancy are showing higher reliabilities although the number of sub-assemblies is increased;
- Floating and moored detachable MECs tend to have larger numbers of sub-assemblies, and therefore a lower predicted reliability, even though they may prove easier to repair;
- Devices that concentrate their complex sub-assemblies on-shore, e.g. Oyster-type WECs, tend to have a higher reliability.

Chapter 7

Availability and its effect on the cost of marine energy

7.1 Introduction and more terminology

Reliability, described in Chapters 3–6, is the probability of a device performing its purpose adequately for the period of time intended under the operating conditions encountered, a definition breaking down into four dependencies:

- Probability;
- Adequate performance;
- Operating conditions;
- Time.

Failure rate, λ, defines $MTBF(\theta) = 1/\lambda$, and is therefore the most essential ingredient for the reliability analysis described in previous chapters. This reliability definition experiences measurement difficulties in continuously operating, repairable systems, such as MECs, because it depends on the four conditions above.

The effect of λ on the operational performance of a continuously operating, repairable MEC, therefore on its Operation & Maintenance (O&M) costs and Operational Expenditure (OPEX), is the device availability, A, the probability of finding a system in the operating state at some time into the future.

This depends upon only two dependencies:

- Operability;
- Time.

Therefore, availability, A, related to $MTTR = 1/\mu$, is a simpler concept for operators than $MTBF = 1/\lambda$ because it has two rather than four dependencies. However, it is much less straightforward to predict during design, because it is more dependent on location, weather conditions and MEC operational support.

A further difficulty with MEC availability is that so few have currently become operational that there is very little operational A data available to investors, designers or operators. This chapter is intended to show how these issues can be considered.

The definition of availability for MECs needs to be clarified and will be based on the availability definition being developed for WTs. Since 2007, an International Electrotechnical Commission working group have been working to produce a standard IEC 61400-Pt 26 to define WT availability, in terms of time and energy output. Until that standard is published, however, there is no internationally agreed definition of WT availability either in terms of time or energy. However, two WT availability definitions, relevant to MECs, have been generally adopted in the United Kingdom and are summarised below:

- **Technical availability**, also known as **system availability**, is the percentage of time that an individual MEC or MEC farm is available to generate electricity, expressed as a percentage of the theoretical maximum;
- **Commercial availability**, also known as turbine availability, is the focus of commercial contracts between MEC farm owners and MEC OEMs to assess the operational performance of a MEC farm project. Some commercial contracts may exclude downtime for agreed items such as requested stops, scheduled repair time, grid faults and severe weather.

In this book, the term availability refers either to technical availability, as defined above, lending itself to the comparison of different MEC farms.

It is fair to say that we know a great deal less about the availability of prospective MEC farms than we do about the potential reliability of MEC components, sub-assemblies, assemblies and systems. This is because availability of data depends upon operational experience, and for MECs that is in very short supply. However, we can make educated guesses about the areas of availability that require most attention. From the above definitions, it follows that technical availability will always be lower than commercial availability, because there is more alleviation of downtime for the former. In addition, an important off-shore issue is that availability, A, is affected by both resource, u, and time, t, $A(u,t)$.

In respect of reliability, the following expressions are useful:

Mean Time To Failure, $MTTF$ (7.1)

Mean Time to Repair, $MTTR$ (7.2)

Logistic Delay Time, LDT (7.3)

Downtime, $MTTR + LDT$ (7.4)

Mean Time Between Failures, $MTBF \gg MTTF$ (7.5)

$MTBF \approx MTTF + MTTR = 1/\lambda$ (7.6)

$MTBF = MTTF + MTTR + LDT$ (7.7)

Failure rate, $\lambda = 1/MTBF$ (7.8)

Repair rate, $\mu = 1/MTTR$ (7.9)

Commercial Availability, $A = (MTBF - MTTR)/MTBF = 1 - \lambda/\mu$ (7.10)

Technical Availability, $A = MTTF/MTBF < 1 - \lambda/\mu$ (7.11)

Note that these are all expressed in terms of variable time, but availability can be expressed in terms of energy production and this may ultimately be more valuable for the operator (Figure 7.1).

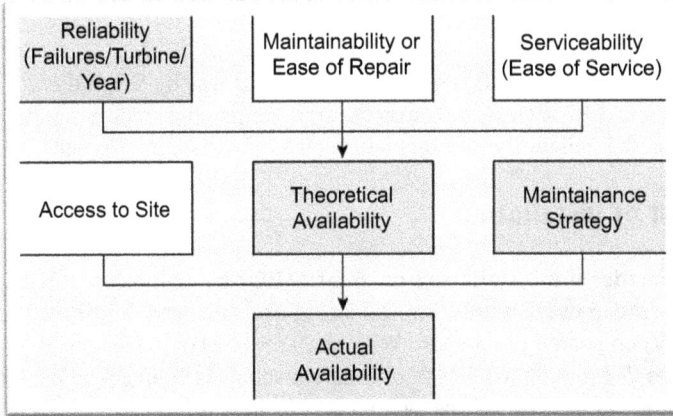

*Figure 7.1 Availability as a function of machine properties, access to site
accessibility and maintenance, Tavner (2012)*

Capacity factor and **specific energy yield** are the commonly used terms to
describe MEC or MEC farm productivity. Capacity factor, C, is defined as the
percentage of the actual annual energy production E (MWh) over the rated
annual energy production, *AEP,* from a MEC or MEC farm of rated power
output, P:

$$C = AEP \times 100/(P \times 8760)\%\tag{7.12}$$

Specific energy yield, S (MWh/m^2/year), is defined as the *AEP* of a MEC,
normalised to its swept device area, A_s (m^2), an approach which is more useful for
TSDs than WECs:

$$S = AEP/A\tag{7.13}$$

The ratio R_S of rated power, $P,$ over the swept device area, A, is a fixed value
for a specific MEC type:

$$R_S = P/A\tag{7.14}$$

Or:

$$R_S = S/(C \times 8760)\tag{7.15}$$

For a specific type of MEC, the specific energy yield, S, is proportional to the
capacity factor:

$$S = R_S \times C \times 8760\tag{7.16}$$

Therefore, the operational performance of a MEC or MEC farm can be defined
as the percentage C or S that expected.

7.2   Previous work

The first published data on European WT reliability was by Schmid *et al.* (1991), a subsequent EU FP7 ReliaWind project, who prepared a report on the previous literature on WT reliability and this has been summarised in Tavner (2012).

7.3   Cost of installation

We can consider the installation costs of MECs by comparison with those of off-shore wind power, whose capital costs are currently estimated at around £1.2 M/MW, compared to on-shore WTs at £0.65 M/MW. Off-shore WT structures are large: the WT hub for a 3.5 MW off-shore machine is 90 m above the sea surface, and the rotor diameter is of the order of 100 m. Initially, the structures will be installed in relatively shallow water depth, 5–20 m, and the weight of each structure will be relatively low, ≈ 400 tonnes, depending on rating.

So, in contrast to typical oil and gas o shorestructures, the applied vertical load to the foundation is relatively small, whereas the wind and wave overturning moment on a MEC structure will be much greater than for an oil and gas structure.

Therefore, an off-shore WT foundation may account for up to 35% of the installed cost. Therefore, off-shore WT and MEC unit capital costs are large and will increase as the MEC farms are placed in deeper water.

However, a single off-shore WT or MEC design can be mass produced for use over a single or many MEC farms, rather than each structure/foundation being individually engineered, as it would be for an oil and gas o shorestructure.

So capital costs of MECs will fall progressively with subsequent projects at later times, and this has been noted in all Danish, Swedish, the United Kingdom, German and Dutch off-shore WT projects.

An interesting comparison in marine renewable capital costs can be made by comparing the experience of two projects in China and the United Kingdom (Figure 7.2):

- China, Shanghai Dong Hai, Da Qiao Bridge off-shore wind farm.
- UK late Round 1 projects.

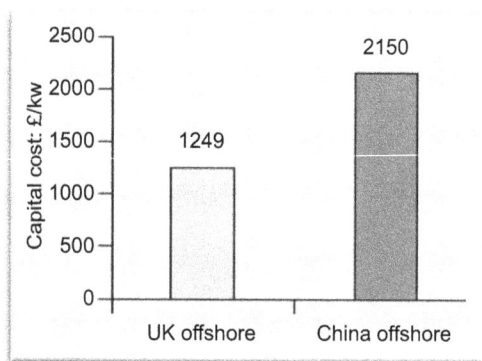

Figure 7.2   Comparison of off-shore wind capital cost between the United Kingdom and China, Tavner (2012)

The capital costs in the Chinese project were £2,150/kW, much greater than those in the United Kingdom at £1,249/kW. This is because China was at the very start of its off-shore development, whereas the United Kingdom has already learnt some valuable lessons. Off-shore wind costs in China will fall as capacity increases, and the same will happen with the deployment of MECs.

Figures for installation costs of MECs are not readily available but Hardisty (2009) gave an indication of prototype and production MEC costs, which are compared with on-shore and off-shore WTs in Table 7.1.

A comparison of the power-to-weight ratio of on-shore and off-shore WTs, TSDs, WECs and floating off-shore WTs, shown in Figure 7.3, gives us a clue that

Table 7.1   Capital cost in £M/MW of prototype and production WTs, TSDs and WECs, based on Hardisty (2009)

Types of machines	On-shore WTs, £M/MW	Off-shore WTs, £M/MW	TSDs, £M/MW	WECs, £M/MW
Prototype	1.4–1.8	2.4–3.0	4.8–8.0	4.0–8.0
Production	0.7–0.9	1.2–1.5	1.4–3.0	1.7–4.3

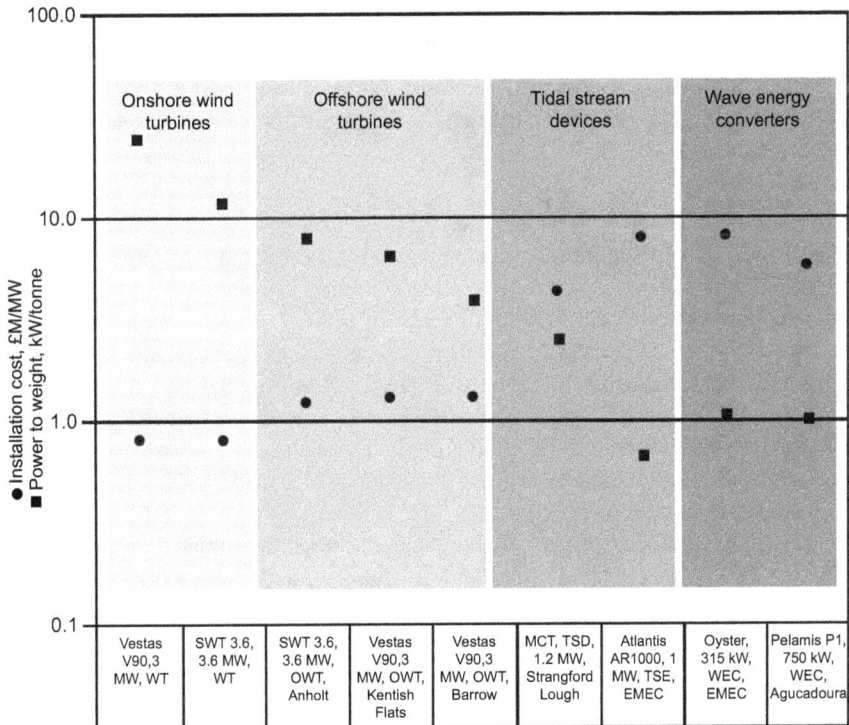

Figure 7.3   Comparison of installation costs and power-to-weight ratios of WTs and various MECs

TSDs and WECs are likely to have £M/MW installation capital costs at least in the range of 3–5 times the figures quoted for off-shore WTs, simply due to the low installed power-to-weight ratios. The figures used in Figure 7.3 were also presented in Figure 1.7.

7.4   Cost of energy

Relatively little can yet be known about the cost of energy (CoE) for MECs, but CoE will commonly be used to evaluate the economic performance of different MEC farms. The following methodology was adopted for wind in a joint report by the International Energy Agency (IEA), the European Organisation for Economic Co-operation and Development and the US Nuclear Energy Agency (NEA). It compared the cost of different electricity production options and a simplified calculation equation was adopted in the United States to calculate the CoE (£/MWh) for a WT system:

$$CoE = (ICC + FCR\ O\&M)/AEP \qquad (7.17)$$

where:
ICC is initial capital cost (£);
FCR is annual fixed charge rate (%);
AEP is annual energy production (MWh);
O&M is annual O&M cost (£).

The result of this approach is the same as that of levelised electricity generation cost used in the wind industry, where the parameter FCR is a function of the discount rate, r, as follows:

$$FCR = r/(1 - (1 + r)^n) \qquad (7.18)$$

where $r \neq 0$. The discount rate r is the sum of inflation and real interest rates. If inflation is ignored, the discount rate equals the interest rate. For the special case of a discount rate $r = 0$, unlikely in the real world, FCR will be ICC divided by the economic lifetime of the MEC farm in years, currently estimated at $n = 20$ years.

A preliminary estimation of the CoE from off-shore wind was carried out by Feng *et al.* (2010) on the early UK Round 1 sites. This shows that at that stage the CoE for off-shore wind in the United Kingdom was about $1.5 \times$ that for on-shore, see Figure 7.4. Improvements in λ and μ cannot yet be made to TSDs, and WECs cannot yet improve on these figures.

The UK subsidised CoE for off-shore wind is therefore estimated, from Round 1, at about £69/MWh against £47/MWh for on-shore wind. An interesting comparison, Figure 7.5, can be made with the CoE for off-shore wind from the Shanghai Dong Hai Bridge project in China, which was ¥980/MWh (i.e. ~£91/MWh), with a project installation cost of ¥23,000/kW (i.e. ~£2,150/kW). Again, one would expect these CoE to fall as experience is gained, the O&M costs to fall and the risks associated

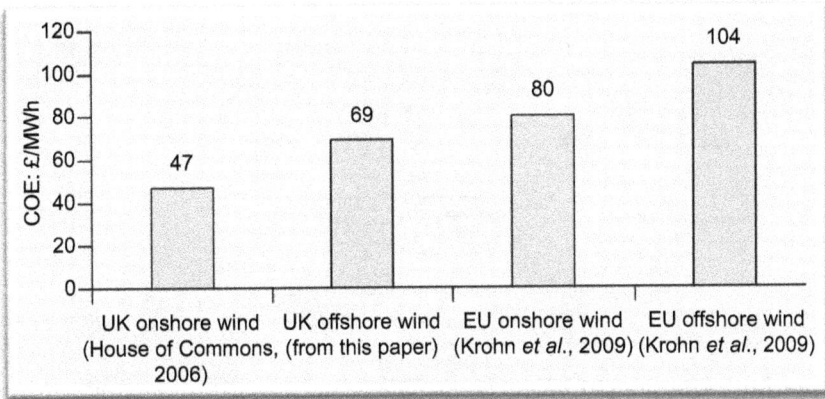

Figure 7.4   Relative CoE for off-shore wind in the United Kingdom and Europe, Tavner (2012)

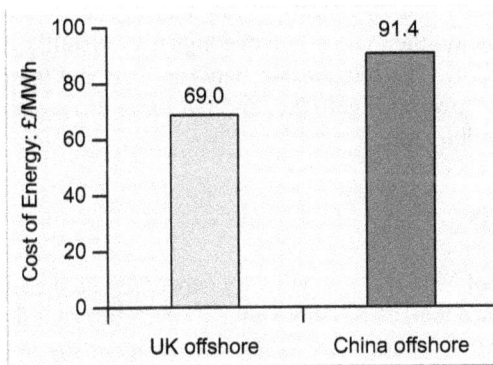

Figure 7.5   Comparison of off-shore wind power cost of energy between the United Kingdom and China

with the capital investment to reduce as has been the experience with European off-shore wind farms.

These calculations were made based on the subsidised CoE, but recent work has stripped away those benefits, showing the true CoE for off-shore wind around the UK coast to be closer to £105–120/MWh. This is falling as experience is gained, capital costs fall and life is extended, the latter being heavily influenced by the O&M regime surrounding the wind farm. There are encouraging signs that, as experience is gathered in off-shore wind, costs are steadily falling towards the current UK target of £100/MWh.

In view of the power-to-weight ratios of Figure 7.3, these figures suggest that MEC CoEs are unlikely to start at less than £350/MWh, and in some cases could be

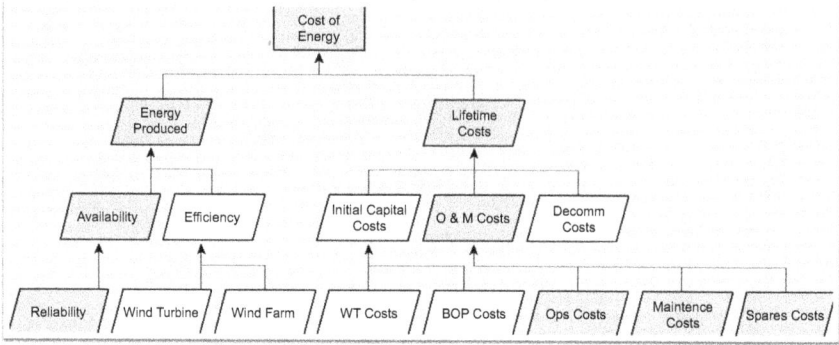

Figure 7.6   Structure of CoE, showing, highlighted in grey, areas of interest for this book

as high as £650/MWh. It will require extensive roll-out and operational experience to reduce these figures.

Early UK Round 1 studies clearly show that operators who impose a high-quality O&M regime on their assets achieve higher availability, lower through-life costs and a lower CoE. The relationship between CoE and the design and operations of the WT is shown in Figure 7.6, and the focus of this book is on the highlighted areas of the diagram.

7.5   O&M costs

The estimated cost of MEC energy will vary depending on the site and project, but Section 7.4 has shown that, as off-shore wind projects are significantly more costly than on-shore, so MEC projects are likely to be more costly than off-shore wind, because the devices are beneath the sea surface.

As MEC designs become adapted to off-shore conditions, the achievement of a favourable economic solution depends upon controlling the MEC farm system's full life-cycle cost. Figure 7.7 illustrates a breakdown of typical total system costs for an off-shore WT farm in shallow water. Much of the price premium now being paid for off-shore wind can be attributed to the WT Foundation, Grid Connection and O&M, in this case offshore approaches 62%. Off-shore conditions require more onerous erection and commissioning operations; meanwhile, accessibility for off-shore routine servicing and maintenance is a major issue.

O&M for off-shore wind farms has proved more complex than on-shore. As a consequence O&M percentage costs for some European off-shore wind farms vary from 18% to 23%, much higher than the measured 12% for on-shore projects. These effects will be similar, if not harder, for MEC farm operators, who can expect O&M costs, as a percentage of the project, much higher than shown in Figure 7.7. During winter, a whole MEC farm may be inaccessible for many days, due to harsh sea, wind or visibility conditions. Even given favourable weather,

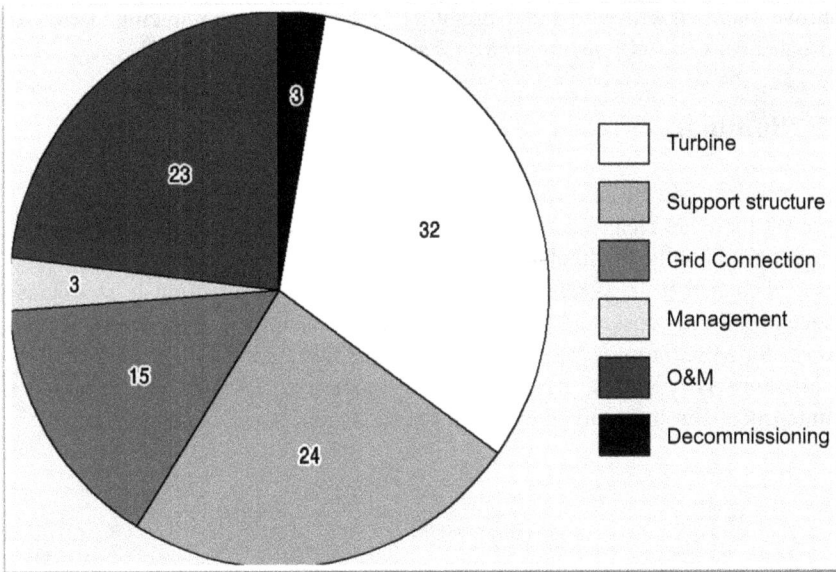

Figure 7.7   Typical cost breakdown for an off-shore wind farm in shallow water, Tavner (2012)

O&M tasks are more costly than on-shore, being influenced by distance off-shore, site exposure, MEC farm size, MEC reliability and maintenance strategy. Off-shore conditions require special lifting equipment to install and change out major sub-assemblies, which may not be available at short notice or be locally sourced. Therefore, advanced techniques are needed to plan maintenance, using data from the monitoring systems fitted to the MECs, requiring a thorough knowledge of off-shore conditions, qualitative physics theory and other design tools to predict failure modes in less conventional ways than has hitherto been done.

Off-shore remote monitoring and visual inspection will become much more important to maintain appropriate WT availability and capacity factor levels, see Chapter 11.

7.6   Effect of reliability, availability and maintenance on cost of energy

Equation (7.17) for CoE can be expressed as a function of λ and μ, allowing us to see the effect of reliability and maintenance on A and CoE as follows:

$$CoE = (ICC \times FCR + O\&M(\lambda,1/\mu))/AEP(A(1/\lambda,\mu) \tag{7.19}$$

Reductions in failure rate, λ, will improve reliability, *MTBF*, $1/\lambda$, and availability, *A,* therefore reducing O&M costs. Reductions in downtime, *MTTR*, will improve maintainability, μ, and availability, A, therefore also reducing O&M. As a consequence, CoE will also reduce as λ and μ improve.

7.7 Summary

Very little can yet be known about the availability of MEC devices and farms; however, there are clear indicators from the on-shore and off-shore wind industries as to how availability is likely to develop.

It is clear that MEC developers need to pay special attention to strategies to reduce installation costs and O&M costs. The next chapter will show that there are choices for MEC farm developers to do this, and these are possible because, unlike an off-shore WT, a MEC device could be moored on location and removed for maintenance. But these involve radical decisions during the development.

Chapter 8

Wave and tidal device layout and grid connection

8.1   Introduction

The reliability of individual MEC devices is actually less important than the overall reliability and availability of the off-shore power stations that they form, and this depends upon the MEC device layouts and grid connections, which will be discussed in this chapter.

We have seen that agreed designs for off-shore wave and tidal arrays, layouts and grid connections have been proposed but not yet achieved, because:

- There is a lack of consensus on the most applicable MEC extraction technologies.
- The need to resolve whether devices should be fixed to the sea-bed or be floating and moored.
- These are dependent on local marine resource, geographic and power network considerations.

The most developed installed MEC arrays are at proving sites, around the world, where arrangements to date have been aimed at technology demonstration, rather than maximising power generation and transmission. Therefore, locations have been based upon the ready availability of existing grid connections, for example at Wave Hub in Cornwall, UK. Greater consideration of connection layout has been taken at some testing stations, for example EMEC, Fall of Warness tidal energy test site, off the island of Eday and Billia Croo wave energy test site, Stromness, both in Orkney, UK, which were designed with electrical connection points available for a variety of devices.

As yet no MEC project is operating a fully developed power collection array, so the following proposals are embryonic. However, off-shore wind has developed some extensive off-shore arrays, the current largest being the London Array off the Thames Estuary, and this array will be used to explain some of the issues involved with regard to reliability and availability.

8.2   MEC arrangements and array cables

Figure 8.1 shows the device and cabling arrangements for two individual MECs, the first fixed to the sea-bed, for example TSDa, fixed monopole device, from

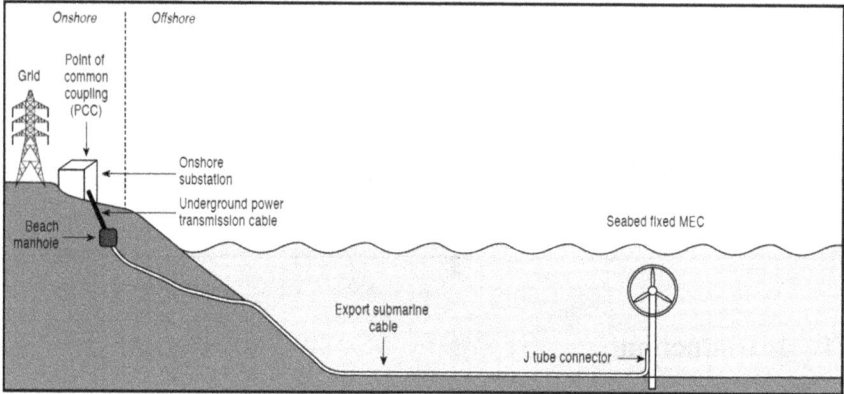

(a)

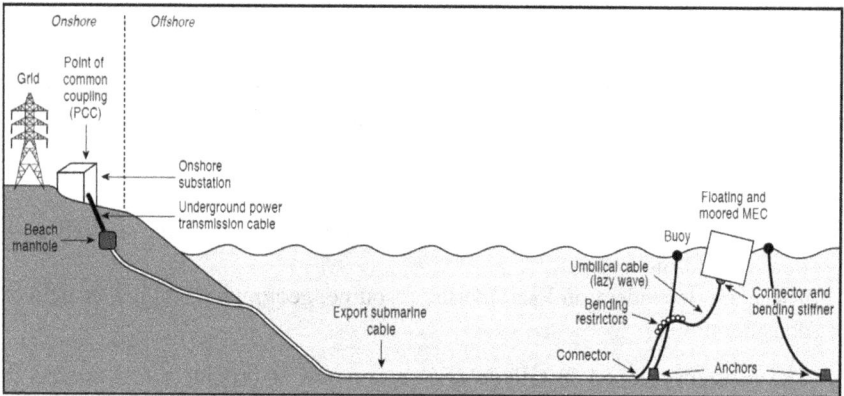

(b)

Figure 8.1   Typical MEC device arrangements, Alcorn et al. (2014): (a) sea-bed fixed MEC and (b) floating and moored MEC

Section 8.6, loosely based on SeaGen, see Figure 6.5, the second floating, for example WECa from Section 6.2, loosely based on Pelamis P2, see Figure 6.4.

The difference between these two arrangements affects the size of the MEC farm and will dictate the cable collection arrangement for the following reasons:

- Fixed MECs can be arranged in a tighter MEC farm as there is no requirement for positional motion of the MEC.
- Floating and moored MECs need to be in a wider arrangement to allow for the spread of mooring anchors and the MEC to adjust itself in the seaway.
- Floating and moored MECs also require greater clearance for shipping, to avoid the risk of damage to mooring lines and flexible cables.

8.3   Device and array layouts

Device and array layouts depend upon the resource and the location. Some prospective MEC farm arrangements, which could be applied to fixed or floating and moored MECs, are illustrated in Figure 8.2 and show the following arrangements:

- **String series cluster**: for larger MEC farms as it offers some redundancy in that not all MECs are dependent on one cable.
- **Full string cluster**: all the MECs are strung on one cable. This offers no redundancy but is cheap and is being adopted for early small MEC farms.
- **Redundant string cluster**: a single loop with switch. This offers some redundancy as it allows MECs to feed on-shore. It is cheap but with the expense of a switch, mounted in a MEC or sub-station.
- **Series DC cluster**: offers extensive redundancy but is designed for MECs generating DC, where generation devices are in series.

The layout for the off-shore WT London Array is shown in Figure 8.3, and is fairly typical of large off-shore wind farms, with a radial structure but separate redundant string clusters, Figure 8.2. Each radial spur is a loop with switches at the end of the loop and at the collector station to provide redundancy in the event of a failure in an array cable. This is a larger example of the third system in Figure 8.2. This array has:

- 175 3.6 MW off-shore WT devices.
- 200 km of fixed, buried, collector XLPE cable array operating at 33 kV AC.
- Collection array connected to two off-shore substations, stepping up voltage from 33/150 kV AC.
- Export from sub-stations via 2 53 km double lengths of fixed, buried, export XLPE cables, operating at 150 kV AC, thus ensuring some redundancy for each sub-station export cable.
- The export cables supply a shore sub-station where connection is made to the 400 kV AC grid via static VAr compensators (SVCs), to balance the reactive VArs produced by the wind farm, and filters to ensure 400 kV waveform compliance.

8.4   AC versus DC connection

Most MECs developed to date generate at LV AC voltage, which allows voltage amplitude to be regulated. This in turn allows devices to be paralleled, voltages to be raised using transformers and transmission to shore at MV AC voltages. It is highly desirable to export power to shore in MV cable, in order to reduce current flow and cheapen cable costs, Figure 8.4(a). Such cables would be trenched and buried but are armoured for protection and sometimes carry the fibre-optic cable for control and SCADA data collection from the off-shore MEC farm.

STRING SERIES CLUSTER IN MEDIUM AND LARGE FARMS (AC AND DC)

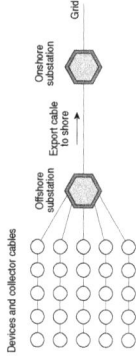

FULL STRING CLUSTER

REDUNDANT STRING CLUSTER

SERIES DC CLUSTER

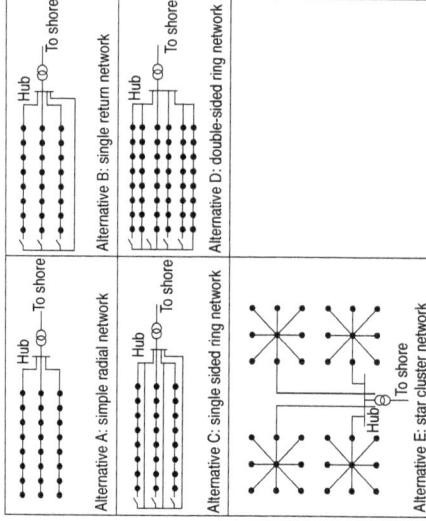

Alternative A: simple radial network

Alternative C: single sided ring network

Alternative E: star cluster network

Alternative B: single return network

Alternative D: double-sided ring network

Figure 8.2 Typical MEC farm layouts, Alcorn et al. (2014)

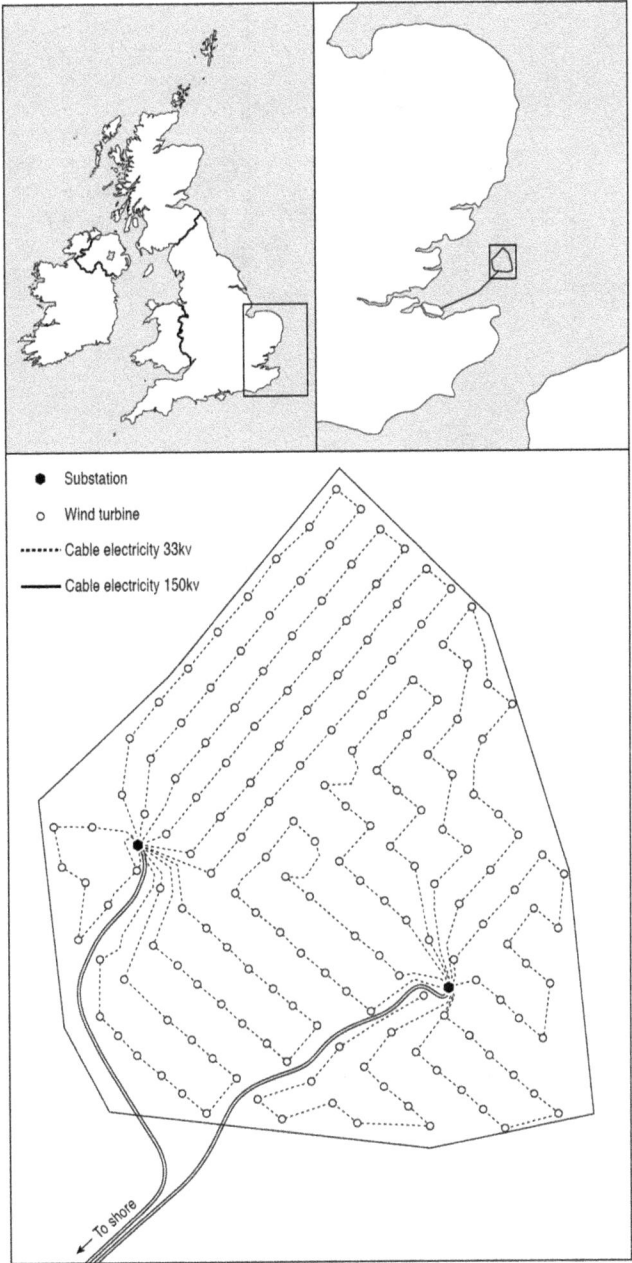

Figure 8.3   Layout of the existing off-shore WT farm London Array

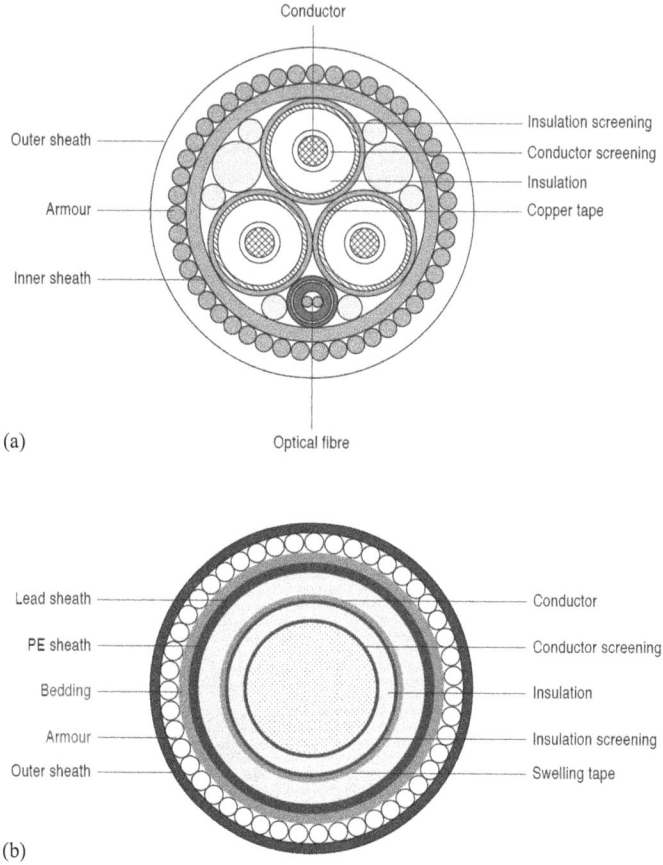

Figure 8.4   Cross-sections of cables from a MEC farm. (a) Three-phase MV AC export cable and (b) single-phase MV DC export cable.

Some smaller MEC devices, however, generate at LV DC, especially those employing permanent magnet generators. This has the disadvantage that voltages cannot simply be raised using transformers but devices can be connected in series to raise voltages, and power be exported to shore using MV DC cables, Figure 8.4(b). This avoids the need to rate electrical equipment for higher currents to cater for reduced power factors, and can allow increased levels of control.

Calculations show that DC transmission over export cables becomes more economic than AC transmission at distances to shore > 40 km (Alcorn *et al.* 2014). But DC operation is still in its infancy off-shore, where control strategies, voltage management and cabling technology are not yet fully agreed, and MEC farms at locations > 40 km from the shore are many years from consideration.

By contrast, AC parallel operation has a long and well-developed experience base, both off-shore and on-shore, enhanced by recent off-shore wind farm operations. It is also entirely suitable for transmission to shore at distances < 40 km, well within the current planned range of WEC and TSD MEC farms.

Figure 8.4 compares the cross-sections of typical AC & DC, and MV export cables.

8.5   Off-shore and on-shore substations

On-shore sub-stations are entirely suitable for small MEC farms of up to 60 MW, with MV AC export transmission, as it avoids the need for large off-shore substation structures as shown in Figure 8.5.

However, as MEC farms increase in size it will become essential to adopt off-shore sub-station structures, such as shown in Figures 8.2, 8.3 and 8.5.

8.6   Effect of array technology and layout on reliability

Despite the small number of MEC production arrays installed and lack of operational experience to date, certain key reliability conclusions can be reached about MEC devices and arrays, based upon the work up to this point in this book:

- Redundancy in wave or tidal MEC cable arrays must raise operational reliability and availability, reducing OPEX and lowering cost of energy. However, there will be a trade-off between OPEX and CAPEX costs for the reliability and availability benefits of that redundancy. Device array and grid connection arrangements will affect this trade-off.
- There is a significant difference in the CAPEX/OPEX ratio between equal area arrays of MEC devices that are floating and moored, and those that are fixed to the sea-bed.
- The geographical area of a MEC array of specified MW rating will be larger for an array of floating and moored devices than for one of fixed devices, because of the larger foot-print of the floating device and moored array. Therefore, the

Figure 8.5   Example of the London Array off-shore sub-station (source: London Array)

annual energy capture in GWh/km^2 and consequent payback will be higher for an array of fixed devices than for a floating and moored device array.

- MECs fixed to the sea-bed are likely to have considerably higher device CAPEX costs than floating and moored devices, because of the higher costs of sea-bed structures.
- MECs with non-detachable connections to fixed cable arrays are likely to have lower array CAPEX costs than MECs with detachable connections, because of the lower cost of detachments and the smaller array area.
- MECs floating and moored are likely to have higher array CAPEX costs than MECs fixed to the sea-bed, because of the higher cost of their larger array area.
- Detachable, floating and moored MECs will be easier to return to dock for maintenance and are likely to have higher availability, lower project risk and lower array maintenance OPEX costs.

Therefore, detachable, floating and moored MECs will probably deliver an OPEX benefit but may carry a CAPEX penalty. However, they may be the best solution necessary to reduce project risk, deliver adequate project operational reliability and pay-off CAPEX costs in an acceptable period. This is demonstrated in Figure 8.6, where the installation costs and prospective CoE are compared

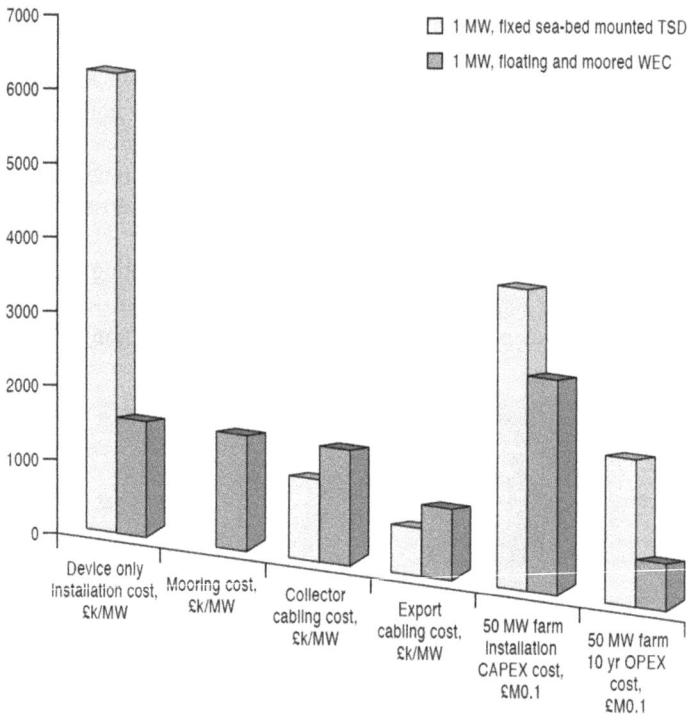

Figure 8.6   Comparison between installation costs, COE, CAPEX and OPEX costs: for a fixed to sea-bed TSD and a detachable, floating and moored WEC

between a fixed sea-bed TSD and a detachable, floating and moored WEC, using data from Hardisty (2009) and O'Connor *et al.* (2012), see discussion in Section 3.6.2. However, such an approach must be treated with extreme caution when considering an immature ocean energy industry, where installation and operational financial data is in extremely short supply.

8.7 Summary

This chapter has shown that there is a balance to be made in device layout and grid connection for MECs that are fixed to the sea-bed and MECs that are detachable, floating and moored.

In general, arrays with floating and moored devices will either contain fewer devices but be of larger geographical area.

There will be an optimisation of the redundancy incorporated into arrays, but experience from off-shore wind farms suggests that the practicable degree of array redundancy will be modest.

There is some interest in the development of DC networks, because of device architecture and possible savings in the transmission of energy from arrays far off-shore. However, experience from off-shore wind farms suggests that long-distance DC connections will be many years away.

There is insufficient installed experience or published information on off-shore MECs to draw specific conclusions about the best way to specify MEC layout and grid connection. However, broad conclusions can be reached about array details:

- Fixed MECs are likely to have a higher unit CAPEX cost and lower array CAPEX cost but higher OPEX costs, than detachable, floating and moored MECs.
- Whereas detachable, floating and moored MEC devices are likely to have a lower unit CAPEX cost and higher array CAPEX cost but lower OPEX costs, than fixed MECs.
- Therefore, detachable, floating and moored MEC devices should deliver an OPEX benefit but could carry a CAPEX penalty. However, they may be the best solution to reduce project risk, deliver adequate project operational reliability and pay-off CAPEX costs in an acceptable period.
- AC collection and export from MEC devices is likely to be the preferred solution for many years to come, before deployment at greater off-shore distances gives advantage to DC devices and DC export.

Chapter 9

Design and testing for wave and tidal devices

9.1 Introduction

The testing required to approve a MEC device for full service is extensive and could take over a decade, as demonstrated by the Pelamis WEC device. However, lessons are being learned and progress to reliable devices is accelerating. It is still important, however, to implement effective testing programmes, and this chapter describes design and testing methods that can be used to drive up the reliability of future MEC devices. Ultimately, it is this that will determine the success of MEC devices in capturing and delivering wave and tidal energy in the long term.

This chapter will review the types of testing of MEC devices which is possible and also desirable.

Some of the descriptions include the prototype testing of very novel devices.

9.2 Design and testing example

An example of the range of small-scale tests necessary for the development of the Pelamis WEC, a successfully MEC device, was given by Cruz, J Ed (2010) and is repeated here in Table 9.1. This shows the extensive use of prominent French, Norwegian and British University test facilities.

This demonstrates the level of testing that may be necessary to deliver reliable MECs, and is based upon extended experience, particularly pioneered by Professor Salter and Edinburgh University. However, lessons have been learnt and the progress to reliable devices seems to be accelerating.

The United Kingdom is also fortunate to have two large renewable energy test facilities operated by the Off-shore Renewable Energy Catapult (ORE Catapult) at Blyth, Northumberland, and at the European Marine Energy Centre (EMEC) in Orkney, Scotland. A summary of their facilities is shown in Table 9.2. Three of the full-size Pelamis devices have been tested at EMEC.

The following sections describe design and testing methods that could be used to drive up the reliability of future MEC devices.

Table 9.1   Scale tests performed on the Pelamis WEC, showing University Laboratories used, Cruz (2010) and EMEC (2013)

Model scale	Test objective	Test location	Facility	Date
1/80	Survivability	University of Edinburgh, UK	Wide Tank	May 1998
1/35	Numerical model validation	University of Edinburgh, UK	Wide Tank	July 1998
1/35	Alternative configurations	University of Edinburgh, UK	Wide Tank	July 1998
1/20	Survivability	City University, London, UK	55 m Wave Flume	Sep 1999
1/20	Numerical model validation	Glasgow University, UK	77 m Wave Tank	Aug 2000
1/20	Power capture and mooring	Trondheim, Norway	Ocean Wave Basin	Oct 2000
1/33	Power capture	University of Edinburgh, UK	Wide Tank	Jan 2001
1/33	Power capture	University of Edinburgh, UK	Wide Tank	Aug 2001
1/7	Digital control systems	Firth of Forth, UK	Sea	Oct 2001
1/33	Mooring response	Glasgow University, UK	77 m Wave Tank	Mar 2002
1/50	Survivability	Glasgow University, UK	77 m Wave Tank	Aug 2002
1/20	Control and survivability	Ecole Centrale de Nantes, France	Wide Tank	Oct 2002
1/20	Control systems	Ecole Centrale de Nantes, France	Wide Tank	Mar 2003
1/7	Mooring response and development	Ecole Centrale de Nantes, France	Wide Tank	Apr 2003
1/1	Pelamis, P1, operational evaluation	EMEC, Billia Croo wave test site	Open Sea	Oct 2004
1/20	Control and survivability	Ecole Centrale de Nantes, France	Wide Tank	Mar 2005
1/21	Control and survivability	Ecole Centrale de Nantes, France	Wide Tank	Feb 2007
1/21	Alternative configurations: numerical model validation	Ecole Centrale de Nantes, France	Wide Tank	Apr 2007
1/1	Pelamis, P2, operational evaluation	EMEC, Billia Croo wave test site	Open Sea	Oct 2010

Table 9.2   Summary of test facilities at Off-shore Renewable Energy Catapult and EMEC

Laboratory	Facilities
Nare, Blyth, Northumberland, UK	Off-shore Wind Turbine Test Field
	Wave test basin with: • Wave flume • Still water tank • Simulated seabed
	50 m Blade Test Facility
	100 m Blade Test Facility
	3 MW Tidal Drive Train Test Facility
	15 MW Wind Turbine Nacelle Test Facility
	Small Wind Test Facility
	Electrical Network Simulation and Test Facility
EMEC, Orkney, Scotland, UK	Billia Croo wave energy test site, Stromness, Mainland Orkney Fall of Warness tidal energy test site, off the island of Eday
	Scale wave test site at Scapa Flow, off St Mary's Bay
	Scale tidal test site at Shapinsay Sound, off Head of Holland
	Office and data facilities in Stromness

9.3   Methods to improve reliability

9.3.1   Reliability results and future devices

The first step to improving reliability is a clear view of the current relative reliability of components in similar and other industries from their measured failure rates, λ, where reliability, $MTBF = 1/\lambda$. as effectively set out in sections in this chapter and Chapter 5.

9.3.2   Design

9.3.2.1   Device design concepts

Based upon the failure rates of components in similar or other industries, a design for any new MEC device should aim to at least match or improve upon those earlier figures.

Particular attention must be paid to the environment from which the measured failure rates have been taken and the environment to which the newly designed component is going, refer to Tables 4.2 and 4.3.

For the marine environment of MECs, the best solution is to bring the component inside a protective enclosure or hull so that a benign environment is achieved. This is standard naval architecture procedure, used in the design of reliable vessels, to ensure sensitive components are enclosed and therefore under environmentally controlled conditions. All off-shore supply vessels take these considerations into account.

Design intent can be achieved either by using identical components or designing improvements which lower the failure rate, using the FMEA or FMECA as a guide to the components requiring attention.

9.3.2.2   Device array design and configuration

Chapter 8 has shown that device array design and configuration, when deploying MECs off-shore, has a significant impact on their potential reliability and therefore on their availability. This is because:

- An array of floating and moored detachable devices will have a larger installed area, km^2/MW, and hence greater access costs, £/MW.
- Whereas an array of fixed devices will have a lower km^2/MW, and hence lower £/MW access costs.
- But fixed and floating detachable devices will require less costly and time consuming marine intervention than fixed devices because they can be detached and maintained at the dock-side, raising their availability.

9.3.2.3   Design review, FMEA and FMECA

A process of regular, documented design reviews for a new device is likely to deliver a reliable design, but design review needs to be driven by some metrics. In the first instance, the device design should be reviewed against the following:

- Its components.
- The design issues of severity associated with those components, as set out in Section 3.2.4 for example:
  - Thermal integrity
  - Electrical integrity
  - Mechanical wear
  - Material integrity
    - Stress
    - Yield
    - Low cycle fatigue
    - High cycle fatigue
  - Corrosion

Failure modes effect analysis (FMEA) can then be used as a valuable design tool to understand the MEC device architecture and focus on its most important components; see Arabian-Hoseynabadi *et al.* (2010) for the application of FMEA to WTs. FMEA is also a powerful design tool for assessing different MEC configurations in terms of the bullets above, as has effectively been done in Chapter 6.

FMEA is particularly useful for considering design improvements for a technology that is changing or increasing in rating, as are MEC configurations. It is a formalised, but subjective analysis, for the systematic identification of possible root causes and failure modes and the estimation of their relative risks. The main goal is to identify and then limit or avoid risk within a design. Hence, the FMEA drives towards higher reliability, higher quality and enhanced safety. It can also be used to assess and optimise maintenance plans. An FMEA is usually carried out by a team consisting of design and maintenance personnel, whose experience includes all the factors to be considered in the analysis.

The causes of failure are said to be root causes, and may be defined as mechanisms that lead to the occurrence of a failure, as described in Section 5.2. While the term failure has been defined, it does not describe the mechanism by which the component has failed. Failure modes, as described in Section 5.2, are the different ways in which a component may fail.

It is vitally important to realise that a failure mode is not the cause of a failure, but the way in which a failure has occurred.

The effects of one failure can frequently be linked to the root causes of another.

FMEA procedure assigns a numerical value to each risk associated with causing a failure, using severity, occurrence and detection as metrics. As the risk increases, the values of the ranking rise. These are then combined into a risk priority number (RPN), which can be used to analyse the system.

$$RPN = S \times O \times D$$

By targeting high value RPNs, the most risky elements of the design can be addressed. RPN is calculated by multiplying the following risk factors:

- Severity, S: referring to the magnitude of the end effect of a failure mode. The more severe the consequence, the higher the severity value assigned to the failure mode. Some authors, including Arabian-Hoseynabadi *et al.* (2010), have linked detection to repair rate, $\mu = MTTR$: a high repair rate indicates a high severity. Therefore, a knowledge of key component repair rates could advise on the value to assign to severity.
- Occurrence, O: referring to the frequency that a root cause is likely to occur, described in a qualitative way. That is not in the form of a period of time but in terms such as 'remote' or 'occasional detection'. Some authors, including Arabian-Hoseynabadi *et al.* (2010), have linked occurrence to failure rate, $\lambda = 1/MTBF$, a high failure rate indicates a high likelihood of occurrence; therefore, a knowledge of key component failure rates could advise on the value to assign to occurrence.
- Detection, D: referring to the likelihood of detecting a root cause before a failure mode can occur.

FMEA is used by various industries, including:

- Automotive
- Aeronautical

- Military
- Nuclear
- Electro-technical

Specific standards have been developed for its application in these various industries.

- SAE J 1739 was developed as an automotive design tool.
- SMC REGULATION 800-31 was developed for aerospace.
- The most widely quoted standard is MIL-STD-1629A (1980).

A typical standard will outline severity, occurrence and detection rating scales, as well as giving examples of an FMEA spreadsheet layout. Also a glossary will be included that defines all the terms used in the FMEA.

The rating scales and the layout of the data can differ between standards, but the processes and definitions remain similar.

With over 50 years usage and development, FMEA has been employed in many different industries for general failure analysis. Due to the complexity and criticality of military systems, it provides a reliable foundation on which to perform FMEAs on a variety of non-military systems, including renewable generation.

Severity, occurrence and detection factors are individually rated using a numerical scale, typically ranging from 1 to 10. These scales, however, can vary in range depending on the FMEA standard being applied. However, for all standards, a high value represents a poor score, for example catastrophically severe, very regular occurrence or impossible to detect. Once a standard is selected, it should be used throughout the FMEA.

In this book, the analyses in Chapter 6 used this standard with some amendment, principally to change the severity, occurrence and detection criteria by which the RPN is calculated. These modifications were necessary to make the FMEA methodology more appropriate to MEC systems.

It can be concluded that the minimum RPN for any root cause is 1, and the maximum is 200. As long as the rating scales of a selected FMEA procedure remain fixed, it can be used for the comparison of alternative designs and identification of critical assemblies. Defining these three criteria tables based on MIL-STD-1629A (1980) is the first step in performing an FMEA. As mentioned before, the basic principles of an FMEA are similar, simple and remain the same, despite differing standards:

- The system to be studied must be broken down into its assemblies.
- Then for each assembly all possible failure modes must be determined.
- The root causes of each failure mode must be determined for each assembly.
- The end effects of each failure mode must be assigned a level of severity, and every root cause must be assigned a level of occurrence and detection.
- Levels of severity, occurrence and detection are multiplied to produce an RPN.

Therefore, the first stage in the FMEA procedure is to obtain a comprehensive understanding of the MEC system and its main assemblies, and the author would suggest that it should be based upon failure rates of similar components, as set out in Chapter 5.

9.3.3   Testing

9.3.3.1   **Sub-assembly and accelerated life testing**

To increase reliability knowledge, the marine renewable industry has been encouraging OEM suppliers to increase the amount of sub-assembly testing. In the wind industry, this has concentrated on:

- Blades
- Gearboxes
- Generators
- Converters

For wave and tidal MECs, sub-assembly testing should also concentrate on the novel sub-assemblies identified in Section 6.4:

- Turbines
- Civil engineering steel or concrete structures
- Monopile or fixed sea-bed location and pinning systems
- Detachable sea-bed location and pinning systems
- Buoyant, hinged steel structures
- Buoyant structure mooring and detachment devices
- LV export cables and detachment devices for buoyant cables
- Hydraulic power units
- Hydraulic receivers, accumulators and tanks
- Variable or fixed speed, pitch-controlled or uncontrolled turbines
- Device controllers and output gain systems
- SCADA monitoring interfaces, sea-shore or shore-sea

Some of this sub-assembly testing is shown in Figures 9.1–9.4.

Accelerated life testing is the process of testing a MEC sub-assembly by subjecting it to stress and strain, temperature and vibration conditions in excess of

Figure 9.1   Tidal turbine blade testing (source: WMC)

Figure 9.2   WT back-to-back gearbox testing (source: ZF Gearboxes)

Figure 9.3   WT brushless DFIG WT generator testing (source: ATB Laurence Scott)

normal service parameters, to uncover faults and potential failure in a short period. By analysing the sub-assembly's response to such tests, designers can predict the service life and maintenance intervals of a product. It should also be possible to use such testing to eliminate premature serial failures (PSF) referred to in Section 3.5.1, Figure 3.20.

Winding insulation and plastic testing can be done at elevated temperatures to produce results more rapidly than could be achieved at ambient temperatures.

So far there appears to be little published material on the accelerated life testing of WT sub-assemblies and none on accelerated life testing of MEC sub-assemblies.

9.3.3.2   On-shore drive train and prototype testing

The testing described in Table 9.1, albeit on a very novel device over a long period of time, shows the level of commitment needed for testing.

Section 9.2 has shown what could be possible on smaller parts of a new MEC design. However, before putting a full-size device into the water, it is necessary to test a complete prototype drive train or generator and gearbox, as demonstrated in Figures 9.5–9.7.

Figure 9.4   WT brushless DFIG WT converter testing (source: Emerson)

Figure 9.5   Fujin Wind Turbine Drive Train Test Facility (source: ORECatapult)

Figure 9.6   Atlantis AR1000, prototype drive train on test (source: ORECatapult)

Figure 9.7   Dynamic subsea cable test rig by OSBIT (source: ORECatapult)

The facilities for such tests are extensive and will be more expensive than the sub-assembly testing described in Section 9.1.3.1, but they will be much cheaper than a full-size prototype installed off-shore. They also offer the potential to gather operational information in an on-shore environment so that the controls and communication for the off-shore MEC can be optimised. They are a valuable precursor to an off-shore production test.

9.3.3.3   Off-shore production testing

Production testing of MEC prototypes at sea has proved costly and has damaged future investment decisions. However, a great deal of off-shore testing of large MECs has been done in a number of countries, and a large amount of operational experience has been gathered; see the examples in Figures 9.8 and 9.9. Perhaps the largest repository of experience is at the EMEC test sites in Orkney.

Figure 9.8   Aquamarine Oyster, prototype on test at EMEC (source: ABB)

Figure 9.9   ScotRenewables SR250, prototype on test at EMEC (source: EMEC)

9.4   From high reliability to high availability

9.4.1   *Relationship of reliability to availability*

The relationship between reliability and availability has been clearly set out in Section 7.1 but can be re-iterated as reliability, $MTBF(\theta)$ having four dependencies:

- Probability
- Adequate performance
- Operating conditions
- Time

    whereas availability, A has only two dependencies:

- Operability
- Time

    Therefore A, related to $MTTR = 1/\mu$, is a simpler concept for operators than reliability, $MTBF = 1/\lambda$, because it has two rather than four dependencies. However, A is much less straight-forward to predict, because it is more dependent on location, weather conditions and MEC operational support, which at this stage in the industry are largely unknown.

    The process of moving from high reliability to high availability is all related to the speed with which MEC defects are detected and how quickly they are rectified, each of these being affected by the subjects of the following paragraphs.

9.4.2   *Off-shore environment*

The environment plays the largest role in this issue and the lack of current operational experience with MECs means that we cannot yet predict the effects of the marine environment on the MECs proposed for mass installation.

9.4.3   *Detection and interpretation*

Detection and interpretation will be thoroughly described in Chapter 10. This will show that there is no doubt that we now have the ability and technology to detect MEC defects very accurately and quickly, based upon very extensive experience in the on-shore and off-shore wind industries. However, MEC designers and developers must take notice of that experience if they are to avoid encountering very serious failures of MEC availability due to poor responses to detectable defects.

9.4.4   *Preventative and corrective maintenance and asset*
management through-life

There are various strategies for dealing with the maintenance of MECs as follows:

- Breakdown maintenance
- Preventative maintenance
- Condition-based maintenance
- Asset management

The unreliability or λ predictions, see Chapter 6, is so high that it will be impracticable to maintain MECs by breakdown maintenance (Wolfram 2006). The consequences of this would be massive mobilisation costs and delays, unacceptable availability, A, as has already been experienced during some MEC off-shore production tests, and high CoE.

Preventative maintenance is practicable but possibly only if floating and moored detachable MECs are deployed, rather than fixed MECs, so that maintenance and repair activities can be concentrated at the dock-side rather than at sea.

Condition-based maintenance is certainly practicable because of the extent and quality of monitoring equipment now available. However, MEC designers and operators must heed the lessons already learnt in other industries, particularly off-shore wind, traditional power generation, railways and aerospace, if they are not to experience expensive repairs and abysmal availability.

The likely method of operating MECs in large farms will be a combination of preventive and condition-based maintenance, organised on an asset management basis. No operator has yet considered this and it must form a part of the plans for deploying MEC farms.

9.5 Summary

The author's simple advice from this chapter is that testing of MECs and their sub-assemblies is essential, but that it must be done before they encounter the harsh off-shore environment.

The costs of undertaking the testing set out in Table 9.1, including off-shore production testing, would be prohibitive for newly developed MECs. Therefore, designers and developers of new devices must organise a smart set of tests that concentrate on the following:

- Evaluation of prospective device reliability, $MTBF = 1/\lambda$.
- Application of FMEA, FMECA design tools for the device, to raise reliability.
- Consideration of the deployment of the device in an array, and the consequent effects on availability A, related to $MTTR = 1/\mu$, in particular, consideration of whether the device should be fixed or floating and moored detachable.
- Consideration of the preventative and condition-based maintenance to be applied to the device, and development of the reliable monitoring necessary to detect prospective failures and support such maintenance.
- Implementation of these systems in a costed, rational, managed test programme including:
  - Sub-assembly and accelerated life testing
  - On-shore drive train and prototype testing
  - Before off-shore production testing.

Chapter 10

Operational experience and lessons learnt

10.1 Introduction

The purpose of this chapter is to present information on operational experience, of renewable MECs. In reality, there is relatively little operational experience with WECs and TSDs, and no published material, since in general only prototypes have been exposed to operational conditions at test or demonstration sites. But there is good experience with wind devices, both on-shore and off-shore, so this will form the basis of the analysis, with a focus on its applicability to MECs. Where information is available for WECs and TSDs, we will elaborate and consider its significance.

10.2 Wind devices in the United Kingdom, on-shore and off-shore

Reliability is a critical factor in the economic success of renewable energy projects. Poor reliability directly affects the project's revenue stream through both increased O&M costs and reduced availability to generate power due to turbine downtime.

The principal objective of reliability analysis is to gain feedback for improving design by identifying weaknesses in parts and sub-assemblies. Reliability studies have also played a key role in optimising WT maintenance strategy. The main factor for optimal preventive maintenance, both from a technical and economical point of view, is the time period selection for inspection. Optimum time period selection cannot be achieved without a comprehensive reliability and availability analysis, resulting in a ranking of critical sub-assemblies. In this way, reliability data can be used to benchmark performance, for organising and planning future O&M, particularly off-shore.

Understanding failure rates and downtimes is difficult, not only because of the considerable range of WT designs and sizes that are now in service worldwide, but also since studies are conducted independently under various operating conditions in different countries.

Whereas the standardisation of data collection practices is well established in the oil and gas industry, the renewable industries have not yet standardised their methods for reliable data collection, see Section 5.3. Based on WMEP and other work, the EU FP7 ReliaWind Consortium proposed a standard approach to WT taxonomy and data collection, catering specifically for larger wind farms (WF), and making use of automatic but filtered Supervisory Control Data Acquisition (SCADA) data and maintainers' logs. This process is set out in Chapter 17. Standardisation of MEC taxonomy and data collection is of paramount importance to facilitate the exchange of information between parties, and to ensure that data can be compared in a useful engineering and management way.

Although modern WTs currently have a shorter design life compared to traditional steam and gas turbine generator systems, i.e. 20 years compared to 40 years respectively, their failure rates have been estimated as about three times those of conventional generation technologies (Spinato *et al.* 2009). Failure rates of 1–3 failure(s)/turbine/year for stoppages ≥ 24 hours are common on-shore (Tavner *et al.* 2007). Despite substantial improvements in recent years, these current figures for reliability of on-shore WTs are still inadequate for the harsher off-shore environment, where maintenance attendance times are increased and availability reduced. For on-shore wind farm sites, high failure rates can be managed by a maintenance regime that provides regular and frequent attendance to WTs. This will be costly or impossible to sustain in remote off-shore sites. Failure rates of 0.5 failure(s)/turbine/year would be desirable off-shore, where planned maintenance visits need to be kept at or below 1 per year.

We are, however, nowhere near this level yet (Tavner *et al.* 2010) and a similar target is needed for TSDs and WECs.

Detailed measurements of failure rates from off-shore WTs have not yet been publicly reported in large enough numbers to give statistically significant results; therefore, the available literature on WT reliability focuses essentially on publicly available on-shore data. WTs constitute a highly specialised technology, and because of the commercial relevance of their failure data, due to the important capital investment and risk of their installation, operators and manufacturers are reluctant to disclose data about reliability or failure patterns. As a result, the sources of information are restricted to a few publicly available databases, although there is a strong argument for WT operators to end this restrictive practice in order to improve economic performance.

Figure 5.2 summarised the analysis of reliability data from three large surveys of European on-shore WTs over 13 years. Failure and downtime data have been categorised by WT sub-assembly. These surveys provide large datasets of failure rate and downtime data, which are remarkably similar, and give valuable insights into the reliability of the various WT drive train sub-assemblies. This data highlights that the highest failure rate sub-assemblies on-shore do not necessarily cause the most downtime. Whilst electrical sub-assemblies appear to have higher failure rates and shorter downtimes, mechanical sub-assemblies, including blades, gearboxes and generator components, tend to have relatively low failure rates but the longest downtimes. The long downtime of the mechanical subassemblies is clearly not due to their intrinsic design weakness but rather the complex logistical and technical repair procedures in the field. It may result from the acquisition time for the spare part, i.e. supply chain, and for the required maintenance equipment. This will be aggravated particularly off-shore where special lifting equipment and vessels are required, and weather conditions have to be considered. In particular, the gearbox exhibits one of the highest downtime per failure among all the on-shore WT sub-assemblies, and is the most critical for WT availability. Also, it has been shown that the replacement of WT major components, such as the gearbox, is responsible for 80% of the cost of corrective maintenance. This suggests that drive train sub-assemblies, such as the generator and gearbox, warrant the most attention.

The datasets shown in Figure 5.9 have some important limitations. Data were taken from greater emphasis on the rotor and power modules because it is believed that these newer variable speed WTs have not yet experienced major gearbox, generator or blade failures to date in service.

A recent study Faulstich *et al.* (2011) has shown that, on-shore, 75% of WT failures are responsible for only 5% of the downtime, whereas only 25% of failures cause 95% of downtime. Downtime on-shore is dominated by a few large faults, many associated with gearboxes, generators and blades, requiring complex and costly repair procedures. The other 75% of faults that cause 5% of the downtime are mostly associated with electrical faults, often caused by the system tripping, which, in the majority of cases, are relatively easy to fix via remote or local resets in the on-shore environment. However, as WTs go off-shore, limited accessibility and longer waiting, travel and work times will amplify the influence of the 75% short-duration failures on off-shore WT availability. Local resets will carry high costs, and difficult access conditions are likely to significantly increase the downtime contribution of these sub-assemblies.

Very little field data is still publicly available for off-shore wind farms, although there are a number of reports published from early publicly funded projects in Europe: Feng *et al.* (2010) and Crabtree *et al.* (2015) carried out a reliability analysis, based on three years of available data from Egmond aan Zee off-shore wind farm in the Netherlands. The wind farm consists of 36 Vestas V90-3MW WTs, situated 10–18 km off-shore and in 17–23 m water depth in the North Sea. Operational reports, available for 2007 to 2009, gave the number of stops resulting from 13 sub-assemblies or tasks, representing 108 WT years of data. The results are shown in Figure 10.2 in the same format as Figure 5.9, but it should be noted that, in this case, stop and not failure frequency was recorded. Therefore, direct comparison cannot be made between these on-shore and off-shore data sets, as these are different concepts. However, the overall distribution is largely similar: sub-assemblies with high stop and failure rates are not always the worst causes of downtime. The control system dominates the number of stops, 36%, but caused only 9.5% of the total downtime. Conversely, the gearbox and generator, respectively, contributed only 6.7% and 2.8% of total stops, but 55% and 15%, respectively, of the downtime. The average energy lost per turbine per year demonstrates similar trends to the downtime (Crabtree *et al.*, 2015).

For large, remote off-shore WT and MEC farms to become cost-effective, improving high availability is essential and requires the adoption of a number of design and operational measures: for example, the choice of the most effective turbine architecture, the installation of effective CM and the application of appropriate O&M programs. Emphasis should be placed on avoiding large maintenance events that require deploying expensive and specialised equipment. However, the most important factor will be to improve the intrinsic reliability of the turbines used, achievable only through close collaboration between manufacturers, operators and research institutes, and the development of a standardised methodology for reliability data collection and analysis, Appendix D.

Despite these issues, Figures 10.1–10.3, taken from Crabtree *et al.* (2015), show the improvement in performance of UK large on-shore and off-shore wind farms over a ten-year period. This indicates the improvements that could be expected from farms of MEC devices, once significant numbers of them are installed.

Figure 10.3 shows the extent of the number of large wind farms in the United Kingdom and the years they have been reporting:

(a) Erected on-shore >100 MW.
(b) Erected off-shore in round 1.
(c) Erected off-shore in round 2.

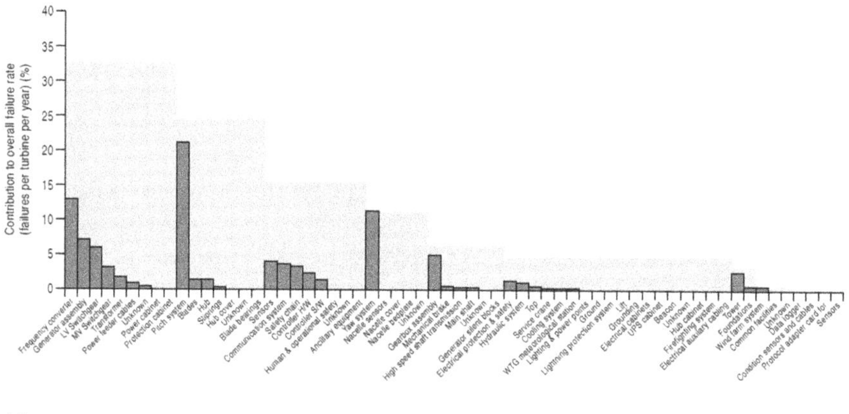

(a)

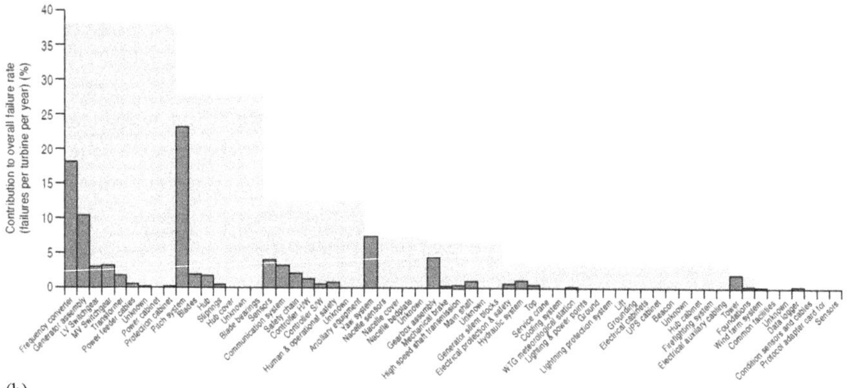

(b)

Figure 10.1   ReliaWind survey failure rate and downtime distributions (Tavner 2012). (a) Sub-assembly failure rate distribution and (b) sub-assembly downtime distribution.

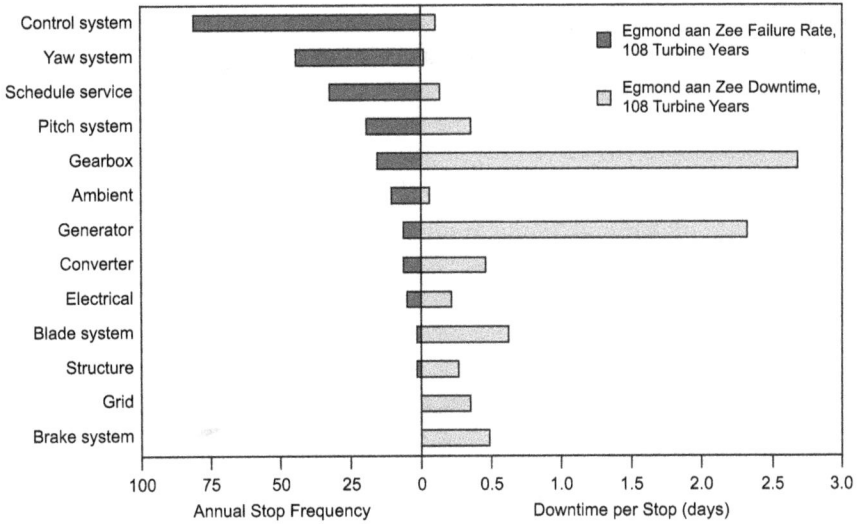

Figure 10.2   Stop rate and downtime data from Egmond aan Zee wind farm over three years (Crabtree et al., 2015)

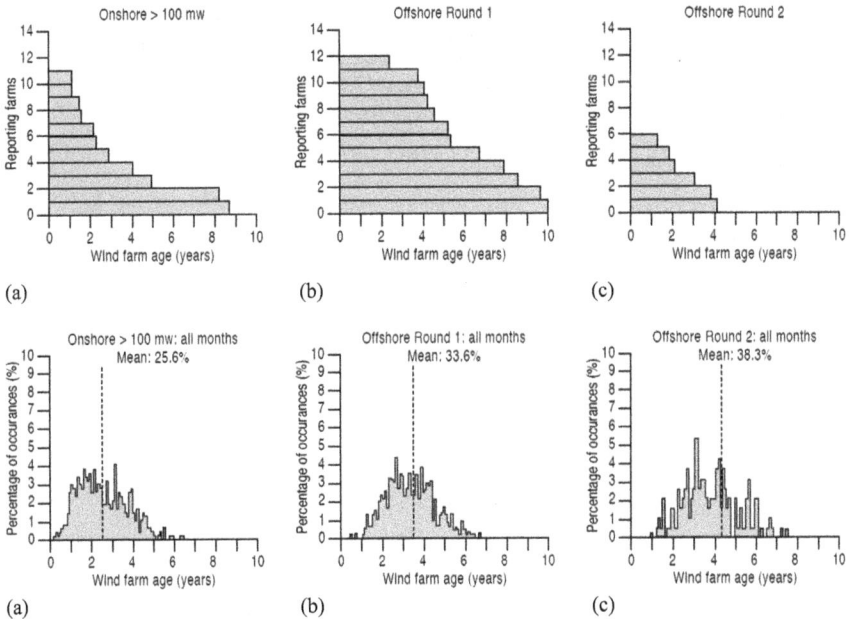

Figure 10.3   Number of reporting wind farms from commissioning and histograms of capacity factors normalised to the total number of reported months (Crabtree et al. 2015). These three figures compare statistical results from operational UK wind turbines: (a) Onshore, (b) Offshore Round 1 and (c) Offshore Round 2.

The number of farm-years for on-shore > 100 MW is relatively small but there are a larger number of off-shore farm-years in round 1 and again, a relatively small number of off-shore in round 2.

Figure 10.4 also shows the resultant mean capacity factors achieved by these off-shore wind farms and this clearly shows the increasing amount of energy converted off-shore and in the larger round 2 site.

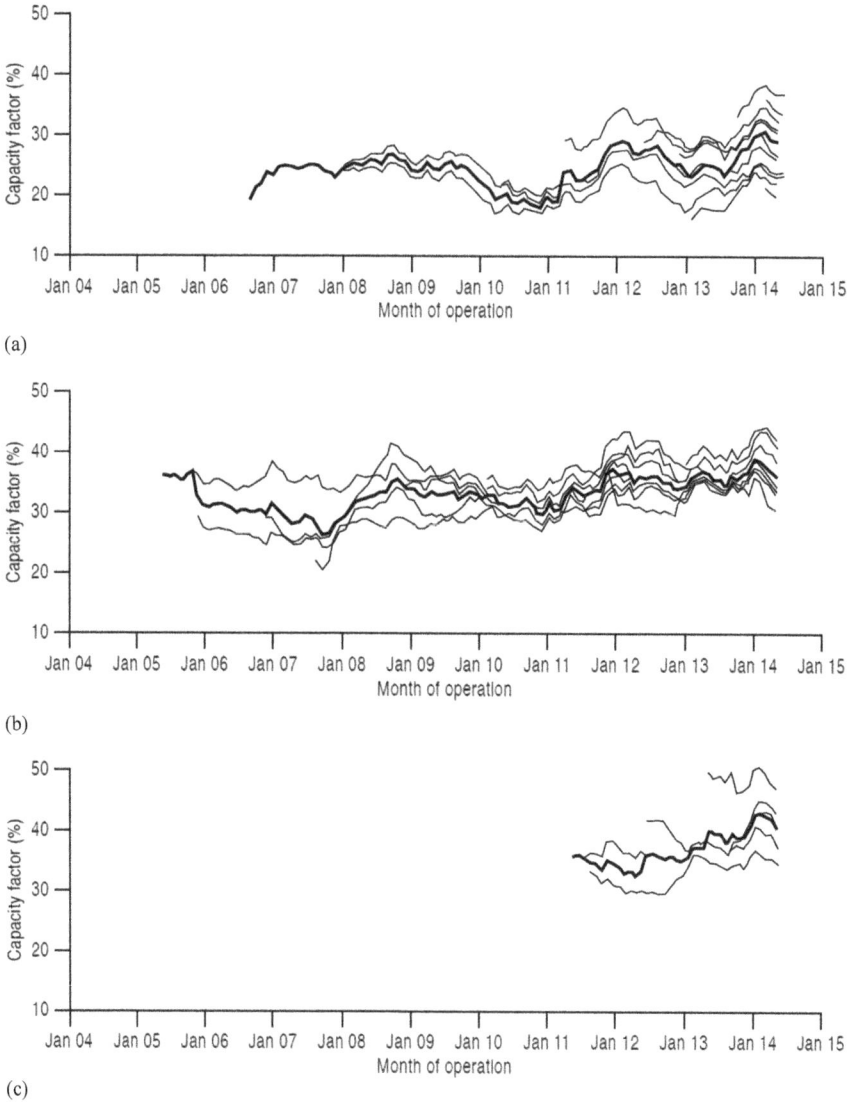

(a)

(b)

(c)

Figure 10.4   Twelve-month moving average monthly capacity factors with overall date for the three populations of wind farms described in Figures 10.1 and 10.2 (Crabtree et al. 2015). These three figures compare time domain results from operational UK wind turbines: (a) Onshore, (b) Offshore Round 1 and (c) Offshore Round 2.

For understanding the prospective future for capacity factor of WEC and TSD farms, Figure 10.4 is most interesting because it demonstrates the progressive improvement in capacity factor, both with time and changing location, on-shore or off-shore. The most striking improvement is with the larger off-shore round 2 sites, implying that the off-shore resource is strong and that, as experience is gained in operating and maintaining these devices, performance improves. However, this is for devices above the sea surface, the critical issue is what can operators expect for WEC and TSD devices that pierce the sea surface?

10.3 Wave devices

There has been no reported operational data from wave devices. The only attempts to relate proposed WEC configuration reliabilities has been reported by Thies (2009).

Delorm (2014) then compared the reliability predictions for WECs and TSDs using the methodology of Chapter 4, and WEC results from Thies (2009). So this was not based upon measured results but prediction, and suggested that we can expect WEC reliability to be worse than TSD reliability. Chapter 8, however, has shown that it must depend upon the design of the device and the decision as to whether to use fixed or floating and moored detachable MECs.

10.4 Tidal devices

The only recorded tidal device data available to the author were reported in 2011 from the Seagen TSD team. This showed an analysis of measured shut-down faults over a period of one year's operation in Strangford Lough, Northern Ireland. The data did not give failure rates, since the number of failures were too few, and the operational period too short, to deliver statistically significant results. However, they did show the distribution of faults, and Delorm (2014) used this to compare with her predicted distribution of faults, based on a reliability model using this device architecture, following the methodology of Chapter 4. That comparison is shown in Figure 10.5.

Figure 10.5 suggests that a reliability prediction model can predict the distribution of failures experienced in service. However, as expressed before, such an approach must be treated with extreme caution when considering an immature ocean energy industry, since measured data was collected from one device, in one location, over a very short period, and the model used was based upon an estimation of device reliability.

10.4.1 Lessons learnt
10.4.1.1 Deployment
Deployment for MECs has been difficult.

There have been many costly deployment failures, particularly for TSDs deployed into an energetic tidal environment.

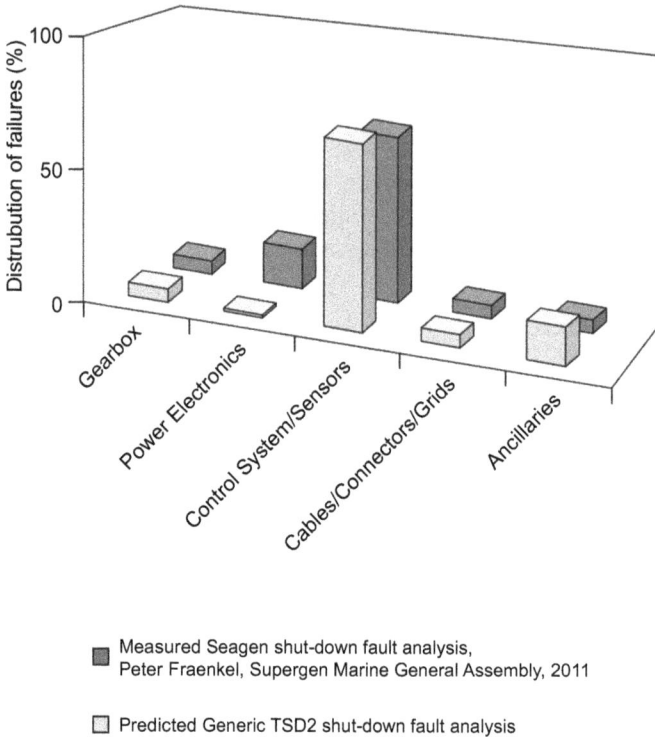

Figure 10.5   Distribution of failures/year: comparison of a TSDa prediction against operational data from SeaGen deployed in Strangford Lough for one year

However, lessons have been learnt and contractors have now refined the techniques. These allow vessels that are dynamically positioned to locate a TSD device over the selected location at slack water and then deploy the device rapidly, with the minimum requirement, for diver intervention.

However, the costliness of such a vessel and the assets needed to locate it accurately, plus the tight timescale for their operation have tended to push some innovators towards floating and moored detachable devices. In that case, the difficulties of location are concentrated into setting anchor points and laying moorings, a well-proven technology for buoys and anchorages.

10.4.2   Device architecture

It is clear from the analysis of Chapter 6 that device architecture determines prospective device reliability, and that there may be great reliability benefit in migrating the generation drive train from the off-shore device to the on-shore connection point. However, this argument weakens as the MEC farm size increases, due to the increase in the complexity of hydraulic connections from sea to shore. Chapter 8 also makes the point that there could be a reliability benefit in developing floating and moored detachable devices, over fixed devices, because of the ease with which the former can be maintained. Although Chapter 6 shows that floating and moored detachable devices have a high sub-assembly count.

However, as has been said before in this book, such an approach must be treated with extreme caution when considering an immature ocean energy industry, where the operational experience of these competing strategies is so small.

10.4.3 Fixed vs floating and moored detachable MECs

The prospective lessons to be learnt from Chapter 8 may be that there could be installation cost, operational availability and cost of energy benefits for floating and moored detachable devices over fixed devices. However, as yet, there is insufficient installed capacity to draw that conclusion with any certainty.

10.4.4 Array configuration

From Chapter 8, it is clear that, for MEC farms, AC will dominate DC connection for many years to come. Experience shows that some cable redundancy must be incorporated in array layouts, but installed off-shore wind farms suggest that the degree of array redundancy, for MEC farms, will be modest.

10.4.5 Corrosion protection

Corrosion protection is concerned with both coatings and active control methods, described by Yebra *et al.* (2010).

10.4.5.1 Coatings

Different coatings are required in each of the following four device areas:

- Topside/ambient/non-immersed surfaces above sea level, continuously exposed to the saliferous atmosphere, requiring conventional marine coatings.
- Wetted/interzone surfaces, continuously exposed to water and the saliferous atmosphere, requiring smooth coatings to slow down the corrosion process.
- Totally submerged surfaces, requiring smooth priming and coating, and the back-up of cathodic protection.
- Moving parts such as turbine blades, buoyant flaps and hydraulic ram structures, requiring anti-fouling coatings.

Again all these systems currently exist for the marine environment and have been successfully piloted in mass-deployed off-shore wind structures.

10.4.5.2   Control

Corrosion control of the submerged metallic parts of the MEC structure will be achieved by using proven cathodic protection systems, already used to protect harbour structures, ship hulls and off-shore WT foundations. These are designed to control corrosion of a metal surface, by making it the cathode of an electrochemical cell. The method of protection is to connect exposed metal parts of the device to more easily corroded, replaceable, sacrifical electrodes, mounted appropriately on the structure and supplied with a controlled DC voltage.

Where metallic surfaces of the device are exposed to sea-water aerated by the operation of a turbine, buoyant flap or buoyant structure, more care must be taken, and more replaceable, sacrifical electrodes need to be mounted close to the area.

10.5   Summary

This chapter has shown that whilst a significant number of TSD and WEC devices have been installed and gained operational experience, none of that experience is visible in the public domain, a point made by Professor Salter in the 1980s and again, more recently, by Cruz (2010).

However, some lessons have been learnt and are widely accepted in the MEC innovator, OEM and developer communities:

- Off-shore wind farm experience provides valuable lessons about the impact of the marine environment on MEC devices, the designs of which must reflect that experience.
- The author strongly believes that the lack of public availability of operational data has diminished the off-shore wind industry's potential to improve operational performance. This should be corrected by agreed industry-wide methods to share reliability and operational data.
- The O&M for off-shore MEC farms is likely to be more arduous and costly than for off-shore wind farms, unless the MEC devices are designed to mitigate O&M costs. Therefore, the same availability of operational data should apply to other MEC technologies if the industry is to grow.
- Off-shore wind experience has shown that progressive cost benefits can be gained in the capacity factor, O&M and CoE, as a result of progressive deployment, and there is no reason to doubt that this will be the case for MEC farms.
- Key issues for MEC innovators, developers and OEMs concerning their devices are:
  - ○ MEC device architecture has a large effect on reliability. OEMs need to incorporate the study of reliability into their designs, and provide redundancy in the drive trains and power electronic conversion, or consider moving these components on-shore.
  - ○ The choice between adopting a fixed or floating and moored detachable MEC will have profound effect on the operational costs of a MEC farm.

There may, therefore, be merit in giving serious consideration to floating and moored detachable MECs.

o   The choice of array configuration can have a significant effect on the installation and O&M costs of a MEC farm that may not be catastrophic but it does require more care than off-shore wind.

o   Corrosion protection is a more significant consideration for off-shore MEC farms and the lessons learnt in large off-shore wind farms are applicable.

Chapter 11

Monitoring and its effect on O&M

11.1 Introduction

Condition monitoring (CM) allows early detection of any degeneration in the components of a renewable generation system, facilitating asset management decisions, minimising downtime, improving availability and maximising productivity. The adoption of cost effective CM techniques has played a role in minimising WT O&M costs for the competitive development of off-shore wind energy. As off-shore WTs operate in remote locations and harsh environments, the need for high reliability and low O&M costs is greater than for on-shore applications. For these reasons, the development of reliable CM, incorporating a number of different systems, for MECs and WTs is essential to avoid catastrophic failures and minimise costly corrective maintenance.

This chapter outlines the current knowledge in the field of CM for renewable generation systems, primarily based on the author and colleague's experience in the electrical machine and wind industry (Coronado *et al.* 2015). Recently published reliability studies are summarised, and the sub-assemblies of most concern for O&M are identified. This is followed by a description of the state-of-the-art in CM, looking at both new and emerging techniques being researched, and industry developed tools, with a particular focus on the economic benefits of CM. Conclusions are then drawn about current systems' challenges and limitations.

The need for successfully detecting incipient faults, before they develop into serious failures, to increase availability and lower cost of energy, has led to the development of a large number of WT monitoring systems. As the wind industry develops, these monitoring systems are slowly being integrated together: this chapter will describe those systems with a view to their application in TSDs and WECs, based on the presentation by Hogg *et al.* (2017).

11.2 Monitoring systems

Modern WTs are equipped with CM systems which allow the active remote control of their functions and may include a variety of systems as follows:

- Supervisory control and data acquisition system (SCADA) monitoring the electronic controller of the device.

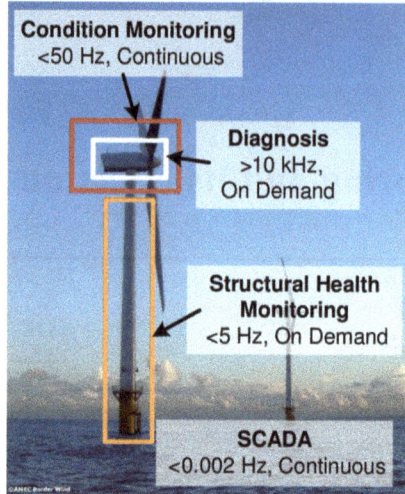

Figure 11.1   Monitoring of a wind turbine (Tavner 2012)

- Condition monitoring system (CMS) for high-risk sub-assemblies such as the drive train or turbine.
- Structural health monitoring (SHM) for the main structure of the device.

The general layout and interaction of the various WT monitoring systems are shown in Figure 11.1.

11.3   Supervisory control and data acquisition (SCADA) system

The SCADA system is a standard installation on large WTs, their data being collected from individual WT controllers. The system provides low-resolution monitoring to inform the WT operation and provide a channel for data and alarms from the WT. The status of the WT and its sub-assemblies is assessed using sensors fitted to the WT. These sensors measure meteorological and turbine operating information at a low sample rate, usually at ten minute intervals in the WT controller. SCADA is a valuable low-cost monitoring system, integrating cheap, high-volume measurement information and communication technology. Data is used to control the WT, and transmitted to a central database for large WT original equipment manufacturers (OEMs). However, wind farm operators rarely use SCADA to monitor WT and wind farm performance. A recent trend in wind project management is to tie the SCADA data into centralised monitoring centres, operated by either the turbine manufacturer or a service provider. This arrangement can provide 24/7 coverage for multiple projects distributed over several time zones. The volume of data from many installations in varied conditions provides better visibility of recurrent problem causes and a synergistic sharing of information.

11.3.1   Commercially available SCADA systems

A survey of the commercial SCADA systems available to the wind industry, conducted by Durham University and the UK SUPERGEN Wind Energy Technologies Consortium, is given by Tavner (2012). This survey provides an up-to-date insight, at the time of writing, into the current state of the art of the commercial SCADA data analysis tools and shows the range of systems currently available to WT manufacturers and operators. The document contains information gathered over several years, through interaction with the SCADA monitoring system and turbine manufacturers. It includes information obtained from various product, and technical brochures and personal interaction with sales and technical personnel at European Wind Energy Association Conferences (EWEA) from 2011 to 2014. The detailed table from the survey, listing up-to-date, commercially available SCADA systems for WTs, from which this summary is derived, is available online. The systems are grouped by monitoring technology and then alphabetised by product name. The survey shows that most of the commercially available SCADA systems are able to analyse real-time data. The WT performance analysis techniques used vary from tailored statistical methods to the use of artificial intelligence.

A number of commercial SCADA systems, such as Enercon's SCADA System and Gamesa WindNet, are customisable Wind Farm Cluster Management Systems, providing, for both individual WTs and wind farms, a framework for data acquisition, remote monitoring, open/closed loop control, alarm management, reporting and analysis, production forecasting and meteorological updates. Five out of the 26 products surveyed, GH SCADA, OneView SCADA system, SgurrTREND, WindHelm Portfolio Manager and Wind Turbine In-Service, were developed by renewable energy consultancies, in collaboration with WT manufacturers, wind farm operators, developers and financiers, to meet the needs of all those involved in wind farm operation, analysis and reporting. Some of the SCADA systems surveyed, such as Alstom WindAccess, claim to feature built-in diagnostics techniques for a timely diagnosis of WT component failures. In these systems, remotely collected data is used to establish benchmarks and identify irregularities, allowing timely intervention to avoid unplanned outages or secondary damage.

Recently proposed SCADA systems, such as the Wind Turbine Prognostics and Health Management platform, developed by the American Center for Intelligent Maintenance Systems (IMS), feature wind turbine modelling for predictive maintenance, and a multi-regime diagnostic and prognostic approach to handle the WTs under various highly dynamic operating conditions.

Some recent SCADA solutions, such as Mita-Teknik Gateway System and ABS Wind Turbine In-Service system, can be adapted and fully integrated with commercially available, conventional, vibration-based CMSs, using standard protocols. These PC-based software packages are designed to collect, handle, analyse and illustrate both operational parameter data from the WT controller, and

CMS vibration signals/spectra, with simple graphics and text. This unified plant operations' view allows a broad and complete analysis of the turbine's conditions, by considering signals of the controller network as well as condition monitoring signals.

In some cases, the SCADA product developer also offers service contracts beyond the manufacturer's service and warranty. They usually include hardware audits, system-specific maintenance plans, monthly checks of the SCADA system, and 24/7 online support. Examples include ABS Consulting's Wind Turbine In-Service and SCADA International's OneView SCADA system.

11.3.2   *Examples of wind turbine monitoring through SCADA systems*

A great benefit of SCADA is that it gives an overview of the whole WT, by providing comprehensive signal information, historical alarms and detailed fault logs, as well as environmental and operational conditions. The main weaknesses of SCADA are that the large volume of data generated requires considerable analysis for online interpretation, and that the low data rate does not allow the depth of analysis usually associated with accurate diagnosis. Potentially, SCADA alarms can help a turbine operator understand the WT and the status of key components, but in a large wind farm, these alarms are currently too frequent for rational analysis. Their added values could be explored in more detail.

Recent research has shown how rigorous analysis of the information collected by SCADA systems can provide long-term fault detection, diagnosis and, in some cases, prognosis, for the main WT drive train sub-assemblies such as the gearbox, converter and pitch system. The implementation of the proposed techniques in the field would result in more intelligent interpretation of SCADA data, enabling automatic WT fault diagnosis and prognosis, giving the operators sufficient time to make more informed decisions regarding asset maintenance.

11.4   Structural health monitoring (SHM)

SHM provides low-resolution signals for the monitoring of key items of the WT structure, tower and foundations. These are particularly important off-shore, where the structures are subjected to strong effects from the sea, wind and seabed. Structural faults are slow to develop and do not need continuous monitoring. SHM systems are frequently installed below the nacelle on large WTs, i.e. >2 MW. Low-frequency sampling signals, below 5 Hz, are recorded from accelerometers, or similar low-frequency transducers, to determine the structural integrity of the WT tower and foundation, which suffers faults driven by blade-passing frequencies, wind gusts and wave slam. Detailed reviews of new and emerging techniques currently being researched in the SHM field are available in the literature.

11.5   Condition monitoring system (CMS)

CMSs provide high-resolution monitoring of high-risk WT sub-assemblies, for the diagnosis and prognosis of faults. Included in this area are blade monitoring systems (BMS), aimed at the early detection of blade defects.

11.5.1   CMS state of the art

Condition monitoring focuses on remotely measuring critical indicators of WT component health and performance, with the objective of identifying incipient failures before catastrophic damage occurs. Failures of the major components of the WT drive train are expensive due to the high costs for the spare parts, the logistic and maintenance equipment, and the energy production losses. A comprehensive online monitoring program provides diagnostic information on the health of the turbine sub-systems, and alerts the maintenance staff to trends that may be developing into failures or critical malfunctions. This information can be used to schedule maintenance tasks or repairs, before the problem escalates and results in a major failure or consequential damage, with resultant downtime and lost revenue. Thus, necessary actions can be planned in time and need not be taken immediately; this factor is of special importance for off-shore plants where bad weather conditions can prevent any repair actions for several weeks. Many faults can be detected while the defective component is still operational. In some cases, remedial action can be planned to mitigate the problem. In other cases, measures can be implemented to track the problem's progression. In the worst case of an impending major failure, CM can assist maintenance staff in logistics planning to optimise manpower and equipment usage, and to minimise the cost of a repair or replacement. In this condition, both the cost of the maintenance activity itself and the costs of production losses may be reduced.

CM has become increasingly important in larger turbines, owing to the greater cost of the components and greater concern about their reliability. Furthermore, it represents an essential technology for off-shore turbines, due to their projected size, limited accessibility and consequent need for greater reliability. The main advantages and benefits arising from applying CMS are summarised in Table 11.1.

The processes necessary for a successful CMS approach are:

- **Detection**: the essential knowledge that a fault condition exists in a machinery component and ideally its location. Without this, no preventive action can be taken to avoid possible system failure.
- **Diagnosis**: the determination of the nature of the fault, including its more precise location. This knowledge can be used to decide the severity of the fault and what preventive or remedial action needs to be taken, if any.
- **Prognosis**: the forecast or prediction of the remaining life or time before failure. Based on this, the most efficient and effective action to remove the fault can be planned. Prognosis has the largest potential payoff of all CM technologies, especially for off-shore turbines.
- **Maintenance action**: repairs or replacements to remove the cause of the fault.

Table 11.1   Characteristics of CMSs

Characteristics	Advantages	Benefits
Early warning	Avoid breakdowns Better planning of maintenance	Avoid repair costs Minimise downtime
Identification of problem	Right service at the right time Minimising unnecessary replacements Problems resolved before the time of guarantee expires	Prolonged lifetime Lowered maintenance costs Quality-controlled operations during time of guarantee
Continuous monitoring	Constant information that the wind power system is working	Security. Less stress

The level of detail required for failure prevention depends very much on the type of component, its perceived value and the consequences of failure. For the main and most expensive WT sub-assemblies, such as the gearbox, the generator and the blades, simple fault detection is not sufficient, as their cost is usually too high to justify total replacement, and some form of diagnosis is required. Diagnosis also allows scope for prognosis: either to predict the possibility of progression from a non-critical fault to a critical fault, indicating the need to closely monitor the fault progression, or to predict the time-to-failure, allowing scheduling of repairs.

11.5.2   Review of CMS techniques

Online monitoring and fault detection are relatively new concepts in the wind industry and are flourishing on a rapid scale. Today, it is common for modern onshore and off-shore WTs to be equipped with some form of CMS. Measurements are recorded from sensors and different methodologies and algorithms have been developed to analyse the data, with the aim of monitoring the WT performance and identifying characteristic fault indicators.

The poor early reliabilities for drive train components led to an emphasis on drive train CMS for WTs. This entered the WT market about 20 years ago, as a result of a series of catastrophic gearbox failures in on-shore turbines, which led to insurers demanding that WT manufacturers take remedial action by utilising CM technology, already applied to other rotating machinery. However, the more recent information on WT reliability and downtime, shown in Tavner (2012), suggests that the target for CM should be widened, from the drive train towards WT electrical and control systems, and blades.

In recent years, efforts have been made to develop efficient and cost-effective CM techniques and signal processing methods for WTs. There have been several reviews on WT CM in the literature, including Yang *et al.* (2012) and (2013), discussing the main CM techniques, the signal processing methods proposed for fault detection and diagnosis, and their applications to wind power.

For advanced CM techniques, signal processing techniques are used to extract features of interest. The selection of appropriate signal processing and data analysis techniques is crucial to renewable device operational success. If fault characteristics can be correctly extracted using these techniques, fault growth can be limited by observing characteristic variations, and provide evidence for fault diagnosis.

Yang *et al.* (2012) provide a detailed summary of the state-of-the-art in WT CMSs, while also providing a comprehensive explanation of the new and emerging techniques currently being researched.

The following monitoring techniques, available from different applications, which are possibly applicable for WTs, have been identified:

- Vibration analysis
- Oil analysis
- Strain measurements
- Thermography
- Acoustic emissions
- Electrical signals

Among these different techniques, vibration analysis and oil monitoring are the most predominantly used for WT applications due to their established successes in other industries.

Vibration analysis is a low-cost and well-proven monitoring technology, typically used to monitor the condition of WT rotating components, i.e. the drive train. Vibration techniques were the first to be used in WT CMS. The principle is based on two basic facts:

- Each component of the drive train has a natural vibration frequency, and its amplitude will remain constant under normal conditions, although varying with drive train speed.
- The vibration signature will change if a component is deteriorating, and the changes will depend on the failure mode.

The type of sensors used essentially depends on the frequency range of interest and the signal level involved. A variety of techniques have been used, including low-frequency accelerometers for the main bearings and higher-frequency accelerometers for gearbox and generator bearings, and in some cases proximeters. By far the most common transducer in use today is the piezoelectric accelerometer, which is applicable to a broad range of frequencies, is inexpensive, robust and available in a wide range of sizes and configurations.

The principles for vibration analysis are presented elsewhere in detail and signal analysis usually requires specialised knowledge. Almost all of the commonly used algorithms can be classified into two categories: time and frequency domains. Time-domain analysis focuses on vibration signal statistical characteristics such as peak level, standard deviation, skewness, kurtosis and crest factor. Frequency-domain analysis uses Fourier methods, usually in the form of a fast Fourier transform (FFT) algorithm, to transform the time-domain signal to the frequency domain. This is the most popular processing technique for vibration analysis.

Further analysis is usually carried out conventionally using vibration amplitude and power spectra. FFT analysers provide constant bandwidth on a linear frequency scale, and, by means of zoom or extended lines of resolution, they also provide very high resolution in any frequency range of interest. This permits early recognition and separation of harmonic patterns or sideband patterns and separation of closely spaced individual components. The advantage of frequency-domain over time-domain analysis is its ability to easily identify and isolate component frequencies of fault importance.

Applying vibration-based CM to WTs presents a few unique challenges. WTs are variable load and speed systems operating under highly dynamic conditions, usually remote from technical support. This results in CM signals that are dependent not only on component integrity but also on operating conditions. One limitation of the conventional fast FFT analysis is its inability to handle non-stationary waveform signals that may not yield accurate and clear component features. To overcome conventional FFT-technique problems and find improved CM solutions, a number of advanced signal processing methods, able to detect defects signatures that are non-cyclic, including time–frequency analysis, bi- or tri-spectrum techniques, wavelet transforms and artificial intelligence, have been developed. These have been researched (Yang *et al.*, 2012, 2013); however, the interpretation of their results is more complex than FFT techniques. Moreover, most new techniques are unsuitable for online CM use because they are computing intensive and have not yet been demonstrated in operating WTs.

Oil analysis is used to determine the chemical properties and content of oil coolant or lubricant with two purposes: safeguarding the oil quality and safeguarding the components involved. Although still expensive, online debris detection in lubricant oil is one of the most promising techniques for use in WT CM. Oil analysis focuses on one of the most critical WT components, the gearbox. Gear wheel and bearing deterioration depends mainly on the lubricant quality, i.e. particle contamination and properties of the oil, and additives used to improve the performance of the oil. Oil monitoring can help detect lubricant, gear and bearing failures, and is an important factor in achieving maximum service life for WT gearboxes. Oil analysis is gradually becoming more important with several on-going pilot projects. Crucial to the value of oil debris detection is the length of the warning that it can give of impending failure, which determines the time to arrange inspection and maintenance. Little or no vibration may be evident while faults are developing, but analysis of the oil can provide early warnings. However, oil debris detection cannot locate a fault, except by distinguishing between the types of debris produced. The combined use of vibration and oil analysis, to cover a broader range of potential failures and to increase the credibility of the CM results, could be a key to WT drive train monitoring.

Strain measurements by fibre-optic sensors are proving to be a valuable technique for measuring blade-root bending moments as an input to advanced pitch controllers. Although they have been demonstrated in operation, they are still too expensive for routine use. However, improvements in costs and reliability are expected. The use of mechanical strain gauges can be useful for lifetime forecasting and protecting against high stress levels, especially in the blades. However, mechanical strain gauges are not robust in the long term, as they are prone to failure under impact and load fatigue. Developments are still needed to improve the instrumentation and sensor robustness, as well as for reliable fault detection algorithms.

Thermography is often used for monitoring electronic and electric components and identifying failure. Infrared thermography is a technique used to capture thermal images of components. Every object emits infrared radiation according to its temperature and its emissivity. The radiation is captured by a thermographic camera. The technique can be applied to equipment from intermittently, but should be done when the equipment is fully loaded, and often involves visual interpretation of hot spots that arise due to bad contact or a system fault. At present the technique is not particularly well established for online CM, but cameras and diagnostic software that are suitable for online process monitoring are starting to become available.

Acoustic emissions could be helpful for detecting drive train, blade or tower defects. Rapid release of strain energy takes place and elastic waves are generated when the structure of a metal is altered. This can be analysed by acoustic emissions. Acoustic monitoring has similarities with vibration monitoring but also a principle difference: whereas vibration sensors are mounted on the component involved, to detect movement, acoustic sensors can attached to the component by flexible glue with low attenuation. For vibration analysis, the frequencies related to the rotational speeds of the components are of interest. For acoustic emission, a wider bandwidth of higher frequencies is considered, which can give an indication of starting defects. These sensors have recently gained much attention, due to their ability to detect early faults, and have been successfully used in research for monitoring bearings and gearboxes. It is considered more effective during low-speed operation of WTs than classic vibration-based methods. However, the acoustic emission technique is still too expensive for routine use, due to the data acquisition costs.

Electrical signals have been widely used for the CM of rotating electric machines and their coupled drive trains (Tavner *et al.* 2008), but have not been used in WTs, due of lack of industry experience. Voltage, current and power measurements, used to control the generator speed and excitation, represent the newest potential source of CM information. The difficulty with these electrical signals is that they are rich in harmonic information, which must be accurately understood if diagnosis is to be performed with confidence. These techniques are at the moment confined to research-related activities, but there is significant potential for applying them successfully in the field. Work has shown the potentiality of the wavelet transform in detecting WT mechanical and electrical faults by electrical power analysis, but this technique still has strong practical limitations because of intensive calculation and consequent inefficiency dealing with lengthy monitoring signals.

Some of the more recent emerging CM techniques described in the research literature include ultrasonic testing, potentially effective for detecting early WT blade or tower defects; shaft torque and torsional vibration measurements, for main shaft and gearbox monitoring; and shock pulse method, an online approach to detecting WT bearing faults.

To date, little work has been done in the area of models to provide prognosis of fault development on the basis of monitoring signals. Much of the research in this area is generic and being conducted by the civil and military aerospace industry. Some specific research will be required to apply the principles of prognosis to the wind power industry. A model for estimating the residual lifetime of generator bearing failure, based on CMS data, has recently been proposed.

11.5.3   Commercially available CMSs

The application of CMS to WTs was requested by the insurance industry in Europe in the 1990s, following a large number of claims triggered by catastrophic gearbox failures. As the drive train is one of the most valuable WT sub-systems, but also most trouble-prone, German insurers introduced this clause as a cost deterrent, to encourage improvements in operating life.

A typical commercially available CMS will feature:

- **Physical measurement**: sensors measure the machine signals: analogue or pulse signals from sensors are then filtered and converted to digital information.
- **Data acquisition system**: data is transmitted from sensors to a processing unit.
- **Feature extraction**: characteristic information is extracted from raw sensor data.
- **Pattern classification**: defect type and severity level are diagnosed.
- **Life prediction**: remaining service life of the monitored component is prognosticated.

Today, a number of commercially certified WT CMSs are available to the wind industry, and they are largely based upon the successful experience of monitoring conventional rotating machines. A survey of the commercially available WT CMSs, conducted by Durham University and the UK SUPERGEN Wind Energy Technologies Consortium, provides an up-to-date insight into the current state CM. It shows the range of systems currently available to WT manufacturers and operators, containing information gathered over several years, through interaction with CMS and turbine manufacturers. It includes information obtained from various product brochures, technical documents and personal interaction with sales and technical personnel at the European Wind Energy Association Conferences from 2008 to 2014. The detailed table from the SUPERGEN Wind survey, listing the up-to-date commercially available CMSs for WTs, from which this summary is derived, is given in Tavner (2012).

The survey shows that the large majority of CMSs currently in use on operational WTs are based on vibration monitoring of the drive train, at a typical sampling rate of up to 20 kHz, with special focus on main bearing, gear teeth and bearings. A vibration-based WT CMS would consist of several sensors and a condition diagnostics system (CDS) enclosure, located in the WT nacelle, and a data server located at the wind farm or a remote monitoring centre. Typically, the system has a channel for the shaft rotational speed, measured by either a dedicated tachometer or the turbine controller. The communication between the WT and the wind farm data server can be through ethernet or fibre optic cables. If no data server is set up at the local wind farm, the system can wirelessly transmit data to a server located remotely, which could be anywhere around the globe. The system normally hosts the CMS software package, which is a platform for reviewing and analysing the data, presenting the CMS results, and streamlining both raw and processed data into a CMS database. One wind farm, typically consisting of hundreds of WTs, can be monitored by one CMS software package. The same systems could be used for MEC farms.

Among vibration-based CMS systems, the main differences are the number of sensors, measurement locations and analysis algorithms, since almost all systems use standard accelerometers as the main physical measurement device. Sensors are mounted on the bearing housing or gearbox casing to detect characteristic vibration signatures for each component. The signature for each gear mesh or bearing is unique and depends on the geometry, load and speed of the components.

According to the survey, the most popular signal processing technique for the vibration monitoring of the WT drive train is the traditional FFT analysis of the high frequency data to detect the fault-specific frequencies. Frequently, time-domain parameters are used to monitor the trend of overall vibration level over time. To minimise data transmission, CMSs analyse data and transmit trends to the system's microprocessor continuously, whereas spectral analysis occurs only when settings detect an unusual condition. A trigger mechanism can be set up in the parameter trending process, based on time-interval or vibration-level information. Whenever it triggers, a discrete frequency analysis snapshot can be taken. Based on these snapshots, detailed examinations of the component's health can be conducted. The amplitude of characteristic frequencies for gears, e.g. meshing frequency, and for bearings, e.g. ball passing frequency, can also be trended over time, to detect potential failures. Such a strategy mitigates the burden of data transmission from the WT; however, it increases the risks of losing raw historic data, because of limited CMS memory size.

In order to acquire data that is directly comparable between each point and, importantly, to allow spectra to be recorded in apparently stationary conditions, a number of commercial CMSs, such as the SKF WindCon 3.0, can be configured to collect the vibration spectra within limited, pre-defined speed and power ranges. This is an important point to note when using traditional signal processing methods, such as the FFT, which requires stationary signals within the analysis time window in order to obtain a clear result.

All vibration-based CMSs surveyed have the capability of carrying out some form of automatic diagnostic procedure. The majority of them are capable of producing alarms, based either on the magnitude of spectral peaks, overall vibration levels or, in some cases, rates of oil debris particle generation. However, results of automatic diagnosis must often be confirmed by vibration experts and component inspection. The level of confidence in these alarms is currently low, but is increasing as monitoring engineers become more familiar with the systems and turbines, and as analysis techniques develop. Automatic diagnosis and prognosis are recent technologies, which still require further investigation.

Only six out of the 27 vibration-based CMSs surveyed state that they are also able to monitor the level of debris particles in the gearbox coolant and lubrication oil system, to enhance their CMS capabilities. Modern oil debris counters take a proportion of the lubrication oil stream and detect and count both ferrous and non-ferrous particles of varying sizes. The counts can be fed as online data to the CMS. However, it should be noted that increasing measurement detail increases the cost of the online instrument.

Some recent commercially available CMSs are beginning to adapt to the WT environment and to be fully integrated into existing SCADA systems using standard protocols. Examples are GE Energy's ADAPT.wind; Romax Technology's InSight Intelligent Diagnostic System; and Gram & Juhl's TCM.

Thanks to this integration, the analysis of the systems installed on WTs can also directly consider any other signals or variables from the entire controller network, for example, current performance and operating condition, without requiring a doubling of the sensor system. The database, integrated into a single unified plant operations' view, allows a trend analysis of the condition of the machine.

Recently, patented condition-based turbine health monitoring systems, such as Brüel&Kjaer's VibroSuite, Romax Technology's InSight Intelligent Diagnostic System and Gram & Juhl's TCM, claim to feature diagnostic and prognostic software, unifying fleet wide CMS and SCADA. This enables the identification of both source and cause of the fault, and the application of prognostics to establish the remaining operational life of the component.

Only three online blade CMSs based on strain measurement using fibre optic transducers have been surveyed. These systems may be fitted to WTs retrospectively. Compared to vibration monitoring techniques, these systems can be operated at low sampling rates, as they are looking to observe changes in the time domain. They are usually integrated in the WT control system but there are also some cases of integration, as an external input, into commercially available conventional vibration-based CMSs. As the blades continue to increase in cost and mass, with the introduction of ever larger wind machines, there is a great deal of concern about their reliability. It is believed that the development of reliable and effective blade monitoring systems will be a key enabler for future megawatt-scale turbines.

No commercial CMS is offered for electronics, beyond oversight by the SCADA system. However, the reliability of both power electronics and electronic controls is of significant concern, especially for off-shore installations where deterioration may be accelerated, in the harsh environment, by corrosion and erosion.

11.6   Cost justification

A CMS based on vibration analysis, such as the SKF WindCon 3.0, is in the range of €15–20k per WT for software, transducers, cabling and installation, more expensive than SCADA with less coverage. The robustness with respect to failure detection/forecasting is not yet completely demonstrated and there has been considerable debate in the industry about the true value of CMS.

The cost justification of CMS for wind power has not been as clear as for traditional fossil-fired or nuclear power plants. To date, there have been few cost evidence publications to support the claims of the CMS industry because of data confidentiality within the industry; however, evidence is accumulating in the off-shore wind industry of the cost benefit of monitoring for asset management in a harsh environment.

These cost-benefit analyses show that lifetime savings can derive from early warning and avoidance of impending failures of critical WT components more than offsetting the lifetime cost of a CM system (Hogg *et al.* 2017). Work has shown that results from a Life-cycle-cost model, evaluated with probabilistic methods and sensitivity analysis, do demonstrate the economic benefit of using CMS in WTs.

The benefit is highly influenced by gearbox reliability. Although the economics of deploying CM for a wind farm is case-dependent, some studies have shown the estimated return on assumed cost being better than 10:1 on-shore, with total return on investment achieved in less than three years. These benefits will be even more dramatic if turbines are installed off-shore where accessibility is a huge challenge.

Yang *et al.* (2013) discuss gearbox failure costs and the cost advantages to be derived from avoiding complete failure through successful WT CM. This was for on-shore and off-shore individual 3 MW turbines and for an off-shore wind farm, Scroby Sands (UK) comprising 30, 2 MW WTs. The work shows that, according to published failure rates, there is clearly a benefit in using WT CMS to eliminate gearbox failures. The figures for Scroby Sands off-shore wind farm are particularly favourable and suggest that off-shore WT CMS is essential for the avoidance of serious downtime and wasted off-shore attendance. In addition, the authors notice that the inclusion of other major sub-assemblies, such as generator, blades and converter, would make a significant contribution to the WT CMS financial case, as would an extension of wind farm working life, but all depends upon the reliability of the WT CMS and the ability of the operator to make use of its indications.

The investment in CM equipment for traditional power generation plants is normally covered by savings of costs from reduced unplanned production losses. For on-shore WTs, unplanned production losses are relatively low. Although for off-shore WTs unplanned production losses are higher, the investment costs for an important part should be paid back by reduction of maintenance cost and reduced costs of increased damage. According to Tavner (2012), WT CMS can only be justified if the system is capable of detecting a fault with early enough warning to avoid full sub-assembly replacement, which is the most costly aspect of failure, and if that CMS detection and warning can be acted upon by operators and WT OEMs.

A practical example of CMS cost justification has been given for an on-shore 25 turbine wind farm operator using the 01dB-Metravib OneProd Wind CMS to successfully detect a broken tooth, on the sun gear of the planetary stage of the gearbox, a generator bearing inner ring defect and a main bearing outer ring defect. This work demonstrated how the CMS investment cost was less than 1% of the price of a current WT, and the benefits coming from 17% failure detection were sufficient to payback it.

11.7   Current limitations and challenges of CM

Experience with CMSs in wind farms, to date, is limited and shows that it is problematic to achieve reliable and cost-effective applications. The application of CM techniques has for decades been an integral part of asset management in other industries, and the technology has in recent years increasingly been adopted by the wind power industry. The general capabilities of the CM technology are, therefore, well known, but the adaptation to the wind industry has proven challenging for successful and reliable diagnostics and prognostics, as they are unmanned and remote power plants. There is still insufficient knowledge among WT maintenance staff of the potential of CMSs, and there is inadequate experience of their application to common WT faults. The main differences to characterise WT operation, compared to other industries, are variable speed operation and the stochastic characteristics of aerodynamic load. This makes it difficult to use traditional frequency-domain signal processing techniques, such as FFT, and to develop effective algorithms for early fault detection and diagnosis, due to the non-stationary signals involved. The majority of commercially available WT CMSs require experienced CM engineers, who are able to successfully detect faults by comparing spectra at specific speeds and loads. Fault detection and diagnosis still require specialised knowledge of signal interpretation to investigate individual turbine behaviour, determine what analyses to perform and interpret the results with increased confidence. The lack of an ideal technique to analyse monitoring signals leads to frequent false alarms, which not only devalue the WT CMS but are also dangerous once a real fault occurs.

One major limitation of the current commercially available CMSs is that very few operators make use of the alarm and the monitoring information available, to manage their maintenance, because of the volume and the complexity of the data. In particular, the frequent false alarms and costly specialist knowledge required for a manual interpretation of the complex vibration data have discouraged WT operators from making wider use of CMSs. This happens despite the fact that these systems are fitted to the majority of large WTs (>1.5 MW) in Europe (Yang *et al.* 2013). Moreover, with the growth of the WT population, especially off-shore, a manual examination and comparison of the CM data will be impractical, unless a simplified monitoring process is introduced. The man-power costs of daily data analysis on an increasingly large WT population will be inappropriate to justify WT CMS.

Most of the WT fault detection algorithms, developed so far, require time-consuming post-processing of monitored signals, which slows down the fault detection and diagnostic process, and still a certain degree of interpretation of the results. These algorithms are generally based on single signal analysis. Diagnosis can be difficult on the basis of a single signal alone, while a multi-parameter approach, based on comparison of independent signals, and able to recognise symptoms from different approaches, has shown increased confidence in the practical applicability of these algorithms, potentially reducing the risk of false alarms.

Aspects of monitoring that particularly concern operators are the improvement of accuracy and reliability of diagnostic decisions, including level of severity evaluation, and the development of reliable and accurate prognostic techniques. It appears evident that cost-effective and reliable CMSs would necessarily require an increasing degree of data automation to deliver actionable recommendations, in order to work effectively in the challenging off-shore environment. The challenge is to achieve detection, diagnosis and prognosis as automatically as possible, to reduce manpower and access costs. Current efforts in the wind CM industry are aimed at automating the data interpretation, and improving the accuracy and the reliability of the diagnostic decisions, especially in the light of impending large-scale, off-shore wind farm generation.

The incorporation of refined, efficient monitoring algorithms and techniques into existing CMSs could represent a way to increase their automation, enhance their capabilities, and simplify and improve the accuracy and user confidence in alarm signals. In particular, automatic processing is required to separate multiple vibration components. These algorithms could reduce the quantity of information that the WT operators must handle, providing improved detection and timely decision-making capabilities. Before being reported to the operators and asset managers, the raw data from remote CM stations would be processed and filtered by an online automatic acquisition system able to:

- Detect incipient faults.
- Diagnose their exact nature and give their location.
- Provide a preliminary malfunction prognosis, through disciplined data management, sophisticated stochastic modelling and computational intelligence, in order to schedule a repair/replacement of the component before failure.

The operator could then choose to examine a particular WT in more detail, if required, so that the costly mobilisation of diagnostic specialists could be minimised only to serious, repairable WT faults. This tool will then allow for a simpler and faster analysis of the different signals transferred by the CMS, since the expert will just receive information relevant to the signals showing defects.

In summary, the main advantages of automatic online CMSs with integrated fault detection algorithms are:

- Appropriate management of wind farm data, reducing human workload without loss of fault detection accuracy, and reducing costs when handling high-rate monitoring information flow from disparate remote locations.
- Elimination of data post-processing.
- Application of multi-parameter approaches, by comparing independent methods contemporaneously, such as vibration, oil debris and electric signatures, to provide more reliable and timely automatic warnings of incipient faults.

- Improvement in accuracy and reliability of diagnostic decisions.
- Improvement in O&M strategy management, according to the automatic prioritisation of fault severity, set by reliable alarms.

Finally, the majority of CMSs currently operate independently from SCADA; therefore, they do not make use of a lot of valuable operational parameter information. It is then expected that the integration of autonomous CM and SCADA systems, within the turbine controller, will lead to more effective monitoring and save costs.

11.8   Examples of monitoring renewable energy sources using CM

Several WT fault detection algorithms for the analysis of discretely sampled CM signals have been proposed in the research literature. The fundamental idea of these algorithms is to reduce the computing demand for standard signal processing by tracking only significant features of WT non-stationary monitoring signals, rather than analysing wide frequency bandwidth signals. The implementation of the proposed algorithms can contribute to automating the fault detection and diagnostic process for the main WT components, improving the confidence in the alarms produced by the system, and reducing uncertainty and risk in applying CM techniques directly to field data from operational WTs.

The influence of rotor electrical asymmetry on stator line current and total power spectra of WT generators has been investigated. The research has been verified using experimental data measured on both Durham and Manchester test rigs, and numerical predictions obtained from a time-stepped electromagnetic model. To give a clear indication of rotor electrical asymmetry in induction machines, a set of concise analytic expressions, describing fault frequency variation with operating speed, are defined and validated by measurement. Simulation and experimental results confirm that analysis of identified fault frequencies in stator line current and power spectra leads to effective generator rotor fault detection and diagnosis. For real-time fault frequency tracking, a Fourier-based algorithm, the iterative localised discrete Fourier transform, $IDFT_{local}$, has been proposed and applied to experimental stator line current and instantaneous power signals, recorded from the Durham test rig, under non-stationary, variable speed, wind-like conditions, as shown in Figures 11.2 and 11.3, respectively (Hogg *et al.* 2017).

When a generator fault was introduced or changed, a step change in $IDFT_{local}$ magnitude was clearly visible.

The results prove that spectral components are valid generator fault indicators under variable load and speed conditions, suggesting that fault severity can be derived from them and demonstrating the effectiveness of the proposed $IDFT_{local}$ algorithm for fault-related analysis in a WT generator.

The results, presented in Figure 11.4, show how a gearbox intermediate speed bearing fault was detected, using multiple signals to raise detection confidence, and using cumulative energy, instead of the conventional signal time axis.

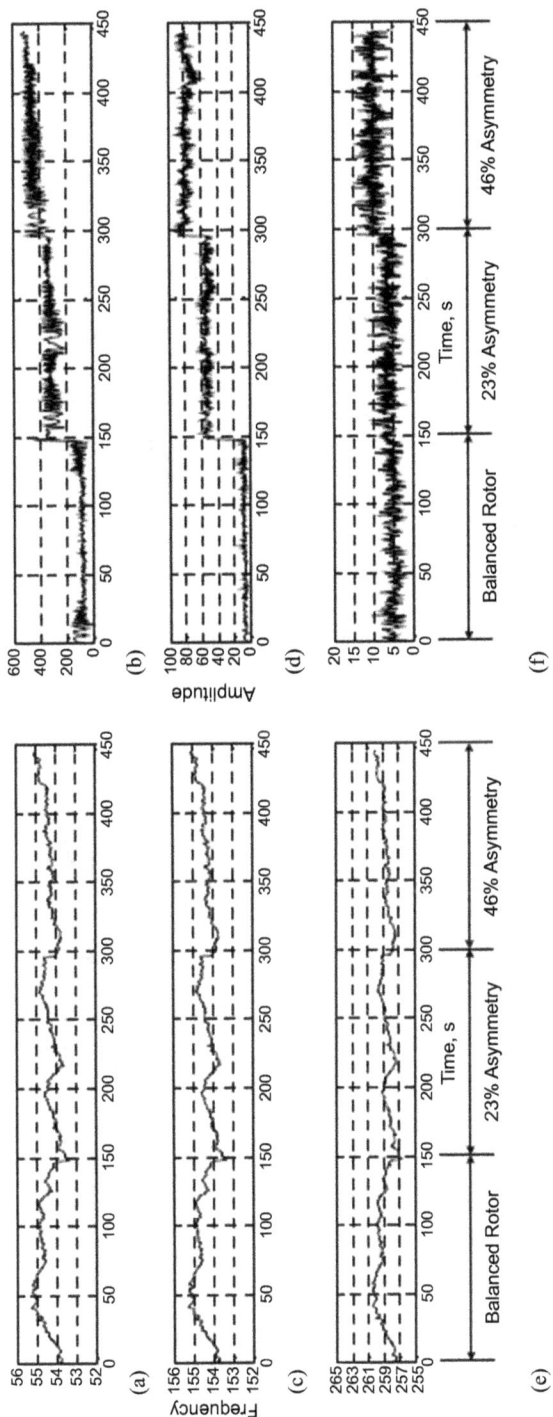

Figure 11.2   Three stator line current IDFT frequency analyses, from no-fault to a progressive drive train fault, using a wind turbine test rig, Hogg et al. (2017)

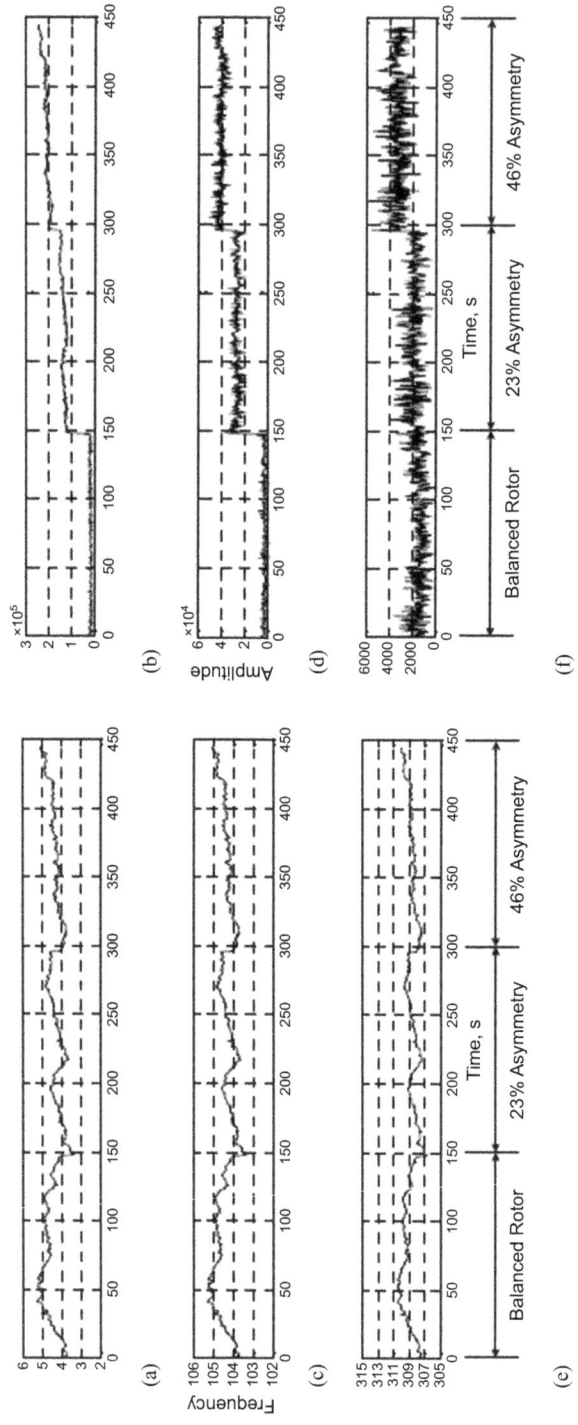

Figure 11.3 Three total instantaneous power IDFT frequency analyses, from no-fault to a progressive drive train fault, using the same wind turbine test rig as in Figure 11.2, Hogg et al. (2017)

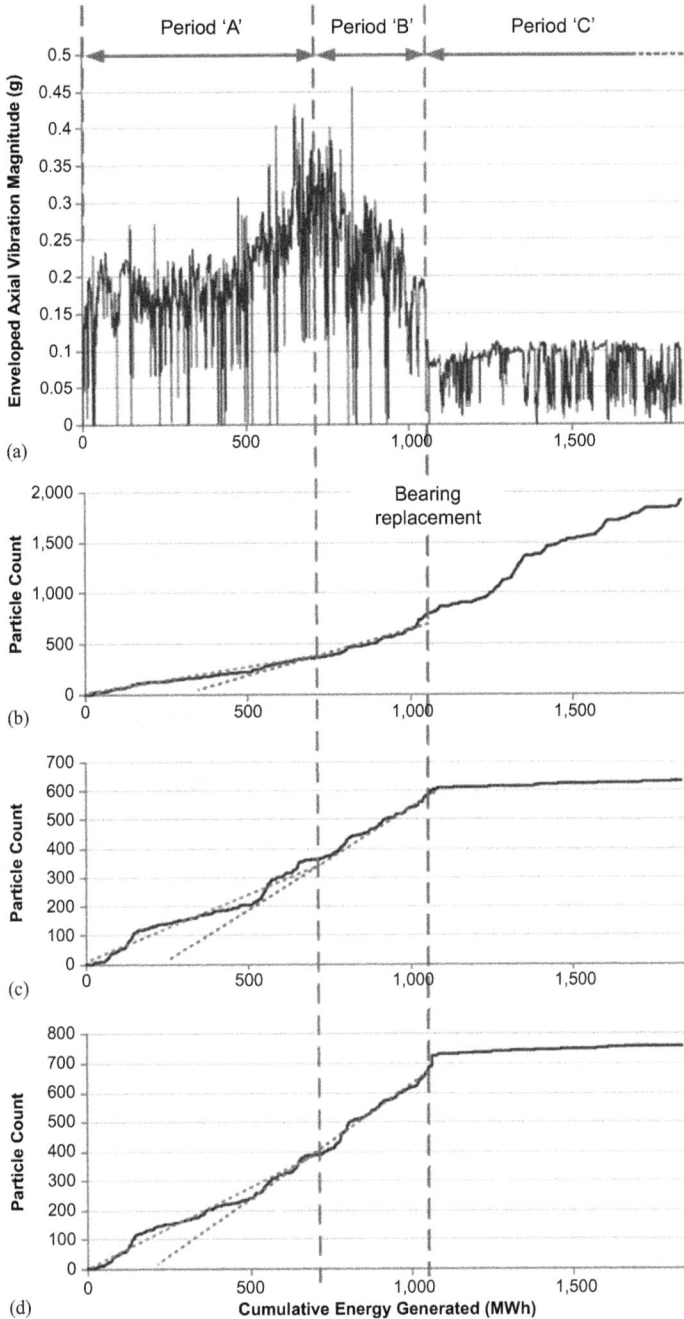

Figure 11.4   *Gearbox condition monitoring vibration and oil analysis plotted against cumulative energy generated (Hogg et al. 2017).*
(a) Gearbox, intermediate-speed bearing, axial vibration magnitude enveloped; (b) cumulative oil debris count for ferrous particles 50–100 μm; (c) 100–200 μm; and (d) 200–400 μm.

During period 'A', an early indication of the incipient fault was provided by increasing independent signals, particle generation rate and vibration magnitude. The diagnosis was then confirmed by the simultaneous change in the two signals when entering period 'B'. During this period, an increasing rate of large particle generation was observed, suggesting significant material breakout from bearing ferrous parts. The corresponding decrease in enveloped vibration suggested a vibration transmission path deterioration, due to breakout of material, a hypothesis subsequently confirmed by visual inspection.

This damage led to the replacement of the gearbox ISS bearing. A significant reduction of vibration envelope magnitude and in 100–200 and 200–400 μm particle generation rate was then observed in period 'C', following maintenance.

However, it is clear from trace b, period 'C' that when the bearing was changed, no attempt was made to change the gearbox oil and remove the damaging metallic particles produced by the replaced damaged bearing, a faulty maintenance action which would almost certainly have led to an early failure of the replacement bearing.

An interesting observation from this series of field measurements was that the operators may have successfully replaced damaged gears.

But they signally failed to filter and clean the gearbox oil, which continued to run, polluted with steel debris produced by the earlier collapsing gear teeth!

An advanced high-sensitivity algorithm, the gear Sideband Power Factor, $SBPF_{gear}$, has also been proposed and is specifically designed to aid automatic online gearbox fault detection, when using field-fitted commercial WT CMS. The uncertainty involved in analysing CM signals is reduced and enhanced detection sensitivity is achieved, by identifying and collating characteristic vibration signal fault frequencies, which can then be tracked as WT speed varies. The influence of the high-speed shaft gearbox pinion fault severity and the variable load operating conditions on the $SBPF_{gear}$ values is investigated by performing both constant and wind-like variable speed tests on the Durham test rig as shown in Figure 11.5. The results show that the proposed technique proves efficient and reliable for detecting both early and final stages of gear tooth damage, with average detection sensitivities of 100% and 320%, respectively. The influence of the fault severity on the $SBPF_{gear}$ detection sensitivity values is evident; the more damaged the pinion, the easier it is to discriminate the fault.

The performance of this experimental technique was also successfully tested, with detection sensitivities of 1,140%–1,251%, on signals from a full-size 750 kW WT gearbox that had experienced severe high-speed shaft gear set scuffing. Once implemented into a WT CMS, the proposed $SBPF_{gear}$ algorithm could facilitate efficient and fast monitoring analysis, reducing each FFT spectrum to only one parameter for each data acquisition.

Generating $SBPF_{gear}$ trends from the vibration spectra, and defining magnitude thresholds for fault severity levels, will indicate to a WT operator when gearbox maintenance action is needed.

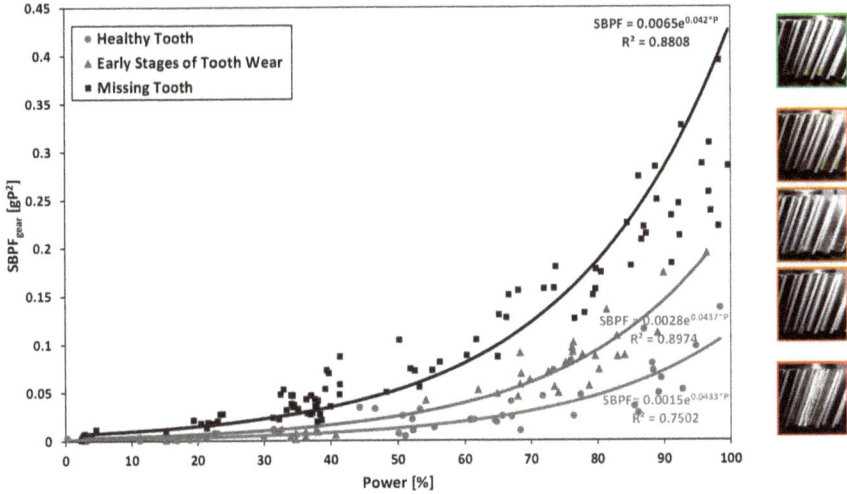

Figure 11.5   Influence of a WT gearbox pinion fault severity under variable load, under progressive seeded-fault tests, as shown in right-hand side photos, Healthy Tooth, Tooth Wear then Missing Tooth Hogg et al. (2017)

11.9   Summary

The evidence shows that monitoring delivers real O&M financial benefits, but operators of renewable devices remain to be convinced.

Part of the reason for this lack of acceptance of monitoring for WTs has been that far too many WT SCADA and CMS signals are monitored.

Operators are then overwhelmed with data and struggle to make sense of the wide variety of signals and consequent analyses.

However, for off-shore MECs, the benefits of monitoring for lowering O&M costs are much greater. Developers of new MEC devices must therefore:

- Analyse their devices to apply only the most useful and necessary monitoring of essential signals.
- Control the outputs of intended monitoring techniques and ensure they can be economically applied to necessary maintenance needs.
- Introduce automatic signal and alarm analysis into any applied SCADA and CMS systems.

Ensure that monitoring of signals and alarms is directly linked to effective maintenance responses that raise MEC availability.

Chapter 12

Overall conclusions

12.1 Staff and training

A mainstream of learning from this book should be that MEC farm OEMs, innovators, designers, developers and operators need some training in predictive reliability estimation for their MECs, if the devices are to be successful in long-term service.

The industry needs greater competence in judging reliability performance between different MEC designs, particularly in considering their maintainability in the field. Some learning from the off-shore wind industry would also be beneficial.

There should be an increase in training, to improve device and prototype sub-assembly testing, particularly based on learning from off-shore wind industry operational experience, where maintenance demands are severe, bearing in mind that MEC operational experience is likely to prove more severe than off-shore wind, particularly for wave devices.

Marine experience in ships and off-shore oil and gas platforms has also shown that the marine environment encourages a high standard of maintenance expertise, which needs to be disseminated from those industries into the new wave and tidal industries.

The reliability and success of our new marine energy generators will depend upon knowledge transfer, exemplified by operational experience such as that shown in Figure 12.1.

12.2 Weather, access and logistics

The weather and logistic issues for off-shore MEC farms, as set out in Chapter 2, have some similarities with those for off-shore wind turbine farms.

However, in the case of MECs, the devices are immersed in the sea and therefore access and maintenance activities inevitably become more complex and require additional techniques and skills to those used by off-shore wind power.

Much has been learnt from off-shore wind operational experience, as shown in Chapter 10. However, it remains to codify experience, ensuring it accurately and appropriately interprets prospective MEC farm industries, as exemplified by Figure 12.2, and where possible maps onto realistic operations.

A beneficial outcome for the new wave and tidal industries would be to develop weather, access and logistic knowledge into training programmes for the

(a)

(b)

Figure 12.1   Off-shore device staff maintenance: (a) tidal device, top-side maintenance and (b) moored device, sub-sea maintenance

Figure 12.2   Floating orbital O2 tidal turbine in operation

TSD and MEC development teams, so that they can understand how to future-proof their technologies, as they come into service.

12.3   Device architectures, reliability and testing

Chapters 3–6 have shown that MEC device architectures will have significant influences on prospective reliability, but design engineers need to be trained in the reliability implications of the MEC architectures they are developing.

Progress in the development of the off-shore renewable industry has been limited by the costliness of prototype deployment and needs to change towards relying more on learnt experience from previous devices, devices from other industries and the modelling and testing prototype sub-assemblies on new designs.

There has to be a way to carry out more extensive simulation, modelling and prototype sub-assembly testing to reduce prototype costs.

The important issue of designing floating and moored detachable MECs rather than fixed devices need to be subjected to analysis to avoid the repetition of costly failures that have occurred on fixed devices early in their service.

Innovators and copycats are the inevitable outcome of rapid technological development and are probably beneficial. An example in the wave energy industry is shown in Figure 12.1(a), the Pelamis wave energy device was developed and tested at the EMEC site in Scotland.

It was interesting to see that a similar device, the Hailong-1, was built tested in 2014 and 2015 in the South China Sea, Figure 12.3.

During the eighteenth and nineteenth centuries similar incidents occurred, for example between the United Kingdom and the United States, as companies started protecting their intellectual property through patents and secrecy.

(a)

(b)

Figure 12.3   Spread of device technologies in different countries: (a) UK 750 kW Pelamis-2 wave MEC device, developed in Scotland, installed and on test in July 2010 at the Billia Croo wave test site Orkney and (b) Hailong-1 wave Chinese wave energy MEC device, 2014 and 2015 tested in the South China Sea, on both occasions rough seas caused tests to be suspended

This is still an important issue and Professor Steven Salter's comment in my Introduction, that we should share information to improve progress, does not take account of the need for emerging companies to protect their intellectual property.

12.4   Maintenance methods and device changeout

The maintenance methods to be used for off-shore MEC farms could be somewhat similar to those currently deployed by the off-shore wind industry.

However, there are challenges to achieve greater than ten years' service.

It will ultimately be necessary for the wave and tidal industries to extend MEC operational lives to 20–30 years.

The results in Chapter 6 clearly show that there may be benefits in deploying floating and moored detachable MECs. These could facilitate device change-out, improving maintenance efficiency.

However, current MECs in development are sea-surface devices, which could limit the occasional maintenance attendance visits that have proved essential in the off-shore wind industry.

The detachment of a MEC from its mooring and returning it to harbour may be essential for major repairs but would be operationally inconvenient for minor maintenance requirements such as oil changes or mandatory visual inspections.

Therefore, radical changes in maintenance methods may be needed if this book's proposal that MEC developers should explore the possibility of deploying floating and moored detachable MECs, with repair activity at a dockside, rather than concentrating on fixed devices with routine maintenance taking place off-shore.

Greater experience of MEC operations needs to be gained with deployed MECs before these issues can be resolved, Figure 12.4.

Figure 12.4   Spread of device renewable ocean energy technologies

12.5   Data management for off-shore asset maintainance

Off-shore WT farm experience, Chapter 11, has shown that these installations have been over-instrumented but under-monitored.

It is clear that much better fault detection methods could be developed, and this could be done by condensing the data collected from existing conventional CMS and SCADA sources and processing them for high-level maintenance management and operational purposes.

This work is still embryonic state and must be resolved for off-shore MEC farms to obtain the benefits of condition-based maintenance. Figure 12.5 gives an example of possible off-shore device data management.

Improvement in this area is parallel to the renewable industry's problems with data sharing. The lack of public availability of operational data has diminished the off-shore wind industry's potential to improve operational performance.

This must be corrected for marine renewables to become viable by agreed industry-wide methods to share reliability and operational data.

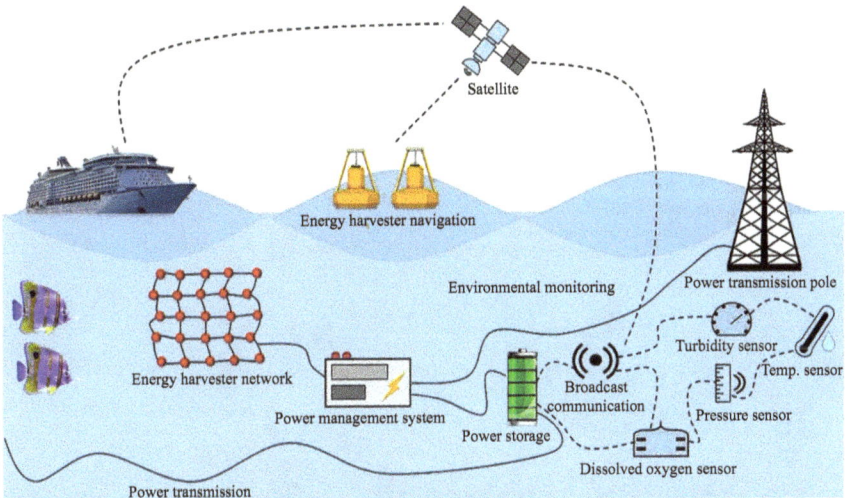

Figure 12.5   Examples of off-shore device data collection

Appendix A
Terminologies

This defines the terminologies used in this book, some adapted from:

- DOE (2011), Reliability Information Analysis Center (2010)
- VGB PowerTech (2007) with some created to satisfy the book content.

Term	Definition
Availability, A	The probability that a machine will be available to operate for a time t and, in this book, is generally quoted as a percentage. A high availability machine spends only short periods of time shut down due to failure or maintenance
Ancillary systems	Systems which are not directly required for the power plant process. This includes heating, ventilation, air-conditioning and monitoring systems
Code letter	Alphabetic character providing classifying information
Device/equipment/ system	A complete piece of machinery able to perform a required function on its own
Failure	When a device fails to perform its energy conversion function. Failure is complete and does not imply partial functionality
Failure mode	The manner in which final failure occurred. For example: • Insulation failure to ground • Structural failure of a shaft
Failure mechanism	The physical manner in which a failure process progresses from the root cause to the failure mode. For example, for the two failure modes above: • Overheating causes degradation of the insulation material, leading to reduced voltage withstand capability • Excessive shock torque on the shaft causing yield and an increase in stress in the remaining parts of the shaft, leading to a progressive and ultimately catastrophic yield of the component
Failure mode and effects analysis, $FMEA$	A subjective analysis tool, defined by standards, that uses a qualitative approach to identify potential failure modes, their root causes and the associated risks in the design, manufacture or operation of a machine

(Continues)

(Continued)

Term	Definition
Failure intensity, $\lambda(t)$	The failure intensity is the rate at which failures occur in a machine, varying according to the operating conditions and age of a machine. It is, therefore, a function of time, sometimes called the Hazard Function. Failure intensity can be constant if machine life is defined. This is termed failure rate, λ, and is, in this book, expressed as failures/machine/year. It is the objective of the maintenance engineer of the machine to keep the failure rate low, constant and predictable
Failure sequence duration	The time from root cause to failure mode. This may be a period of seconds, minutes, days, months or weeks, and can depend on: • The failure mode itself • The operating conditions of the machine • The ambient conditions
Mean time to failure, *MTTF*	The average mean *TTF* of successive failures (see definition for *TTF*)
Mean time to repair, *MTTR*	The mean *TTR* of successive failures and can be derived from a number of machines in a population (see definition for *TTR*)
Mean time between failures, *MTBF*, θ	The mean *TBF* of successive failures and can be derived from a number of parts in a population (see definition for TBF). *MTBF* is the sum of the *MTTF* and *MTTR*
Mission time, mission reliability	The operational time in service of an MEC fulfilling its generation mission. The total amount of mission time, divided by the total number of critical failures during a stated series of missions (MIL-STD-721B)
Part	Entity treated in the process of design, engineering, operation, maintenance and demolition. Could be a system, sub-system, assembly, sub-assembly or component of a WEC or TSD
Random failures	Failures that occur unpredictably during the useful life period of a system, WEC or TSD
Reciprocating device	A device that uses rectilinear motion. For example, an oscillating hydrofoil uses the flow of water to produce the lift or drag of an oscillating part transverse to the flow direction. This behavior can be induced by a vortex, the Magnus effect or flow flutter. Mechanical energy from this oscillation can then convert mechanical energy, via a linear or rotating electrical machine, to electrical energy
Redundancy	The existence of one or more engineering means, not necessarily identical, for accomplishing one or more function of a system: • **Active redundancy** has all sub-systems operating simultaneously • **Standby redundancy** has alternate means activated upon failure of a sub-system

(Continues)

(Continued)

Term	Definition
Reliability, *R*	• The probability that a part can operate without failure for a time, *t*, and in this book is generally quoted as a probability percentage • The duration or probability of failure-free performance under stated conditions • The probability that a part can perform its intended function for a specified interval under stated conditions. For non-redundant parts, this is equivalent to the first definition. For redundant parts, this is equivalent to the MIL-STD-721B definition of mission reliability Reliability as a function *R(t)* is sometimes known as the survivor function because it indicates what proportion of the starting population survives at a particular time, *t* A high reliability part has a high MTBF, a high percentage reliability and a low failure rate
Reliability model	A model identifying the reliability of a system. A reliability model integrates the interrelation of parts for reliability analysis and assessment
Reliability prediction	A measure for estimation of figures of merit for product reliability performance
Root cause	The manner in which a failure mode was initiated. For example for the two failure mode cases given above: • Overheating of the insulation could be the root cause leading to the failure mode of insulation failure to ground • Excessive shock torque being applied to the shaft could be the root cause leading to the failure mode of structural failure of a shaft
Root cause analysis, *RCA*	A method of problem-solving used, following failure, to identify the failure modes and underlying root causes
Rotating device	A device that uses rotating motion. For example, a turbine which can convert mechanical energy via a rotating electrical machine to electrical energy
Configuration or taxonomy	The configuration of a WEC or TSD can be described as a taxonomy. This book will use the following definitions:
	System	A set of interrelated parts, e.g. an MEC, which will be made up of sub-systems, assemblies, sub-assemblies and components
	Sub-System	A part within such a TSD or WEC system, e.g. a drive train
	Assembly	A part within such a sub-system, e.g. a gearbox
	Sub-assembly	A part within such an assembly, e.g. a gearbox lubrication motor and pump
	Component	An individual part of such a sub-assembly, e.g. a lubrication motor bearing

(Continues)

(*Continued*)

Term	Definition
Time to failure, *TTF*	The time measured from the instant of installation of the part to the instant of failure, usually given in hours. It does not include the time to repair, as a result of failure
Time to repair, *TTR*	The time measured from instant of first failure to the instant when the machine is available for operation again, usually given in hours
Time between failures, *TBF*	Under the hypothesis of minimal repair, that is repair that brings a part back to its condition before failure, *TBF* is the time measured from the instant of installation of part to the instant after the first failure when the part is available for operation again. TBF is usually given in hours

Appendix B
Device configuration

To aid the description of device taxonomies, the German Power Standards Company VGB Nomenclature has been used wherever possible.

An indicative terminology for a tidal stream device (TSD) as used in this book letters is shown below with useful identifying code/lettering:

			Code/description	
AA	Grid connection sub-system			
	AAG	Grid connection assemblies		
		AAG10	LV load-break switch system	
		AAG20	LV system black start	
			AAG21	Converter AC/DC
			AAG22	LV battery
			AAG23	Control and management sub-system
		AAG30	Transformer assemblies	
			AAG31	Transformer sub-assembly
			AAG32	Transformer cooling sub-assembly
			AAG33	Main circuit-breaker 11 kV

(Continues)

(*Continued*)

				Code/description		
				AAG33/QA001	Transformer isolator switch
		AAG40	Transmission cable		
AB	Corrosion protection sub-system, corrosion control				
B	LV DC uninterrupted electrical supply sub-system				
	BU	LV electrical assemblies			
		BUU	LV electrical sub-assemblies		
			BUU10	DC/AC converter	
			BUU11	400 V LV supply isolator	
			BUV10	Battery	
CA	Control and management sub-systems				
	CA10	SCADA system			
	CA20	System controllers			
		CA21	HVAC controller		
		CA22/ MDX11	Hydraulic controller		
		CA23/ MKA11	Generator automatic voltage regulator		
		CA24/ MKY11	Converter controller		

(*Continues*)

(*Continued*)

Code/description			
	CA25/ XAA30	Environmental controller	
DT Drive train sub-systems			
MD Turbine sub-system			
MDA	Turbine assemblies		
	MDA10	Rotor blades	
	MDA20	Hub	
MDC	Pitch assemblies		
	MDC10	Pitch systems	
	MDC11/UP001	Pitch bearings	
MDK	Shaft assemblies		
	MDK10	Main shaft	
	MDK11/UP001	Main bearing	
	MDK11	Shaft seals	
	MDK20	Gearbox	
	MDK21/UP001		Gearbox bearings
	MDK22		Gearbox seals
	MDK30	Brake	
	MDK40	Couplings	
MDV	Lubrication and cooling assemblies		
	MDV10	Lubrication oil cooler	
	MDV20/ MDX30	Hydraulic oil cooler	
MDX	Hydraulic assemblies		
	MDX10	Hydraulic power pack	
	MDX20	Hydraulic hoses	

(*Continues*)

(*Continued*)

			Code/description	
MK Generator sub-systems				
	MKA	Generator assemblies		
		MKA10	Generator sub-assembly	
	MKC	Generator protection assemblies		
		MKC10	Generator circuit-breaker, 400 V	
		MKC11	Generator protection unit	
			MKC20 QA001	Generator isolator switch
	MKY	Converter assemblies		
		MKY10	AC/AC converter	
U Structure, nacelle/ foundation/ moorings				
X Ancillary systems				
	XA	Ancillary sub-systems		
		XAA	Environmental sub-systems	
			XAA10	Ventilation system
			XAA20	Water-cooled heat exchanger

Appendix C

MEC reliability diagrams

Six MEC device reliability has been prepared as examples as follows:

C.1 Wave energy converters

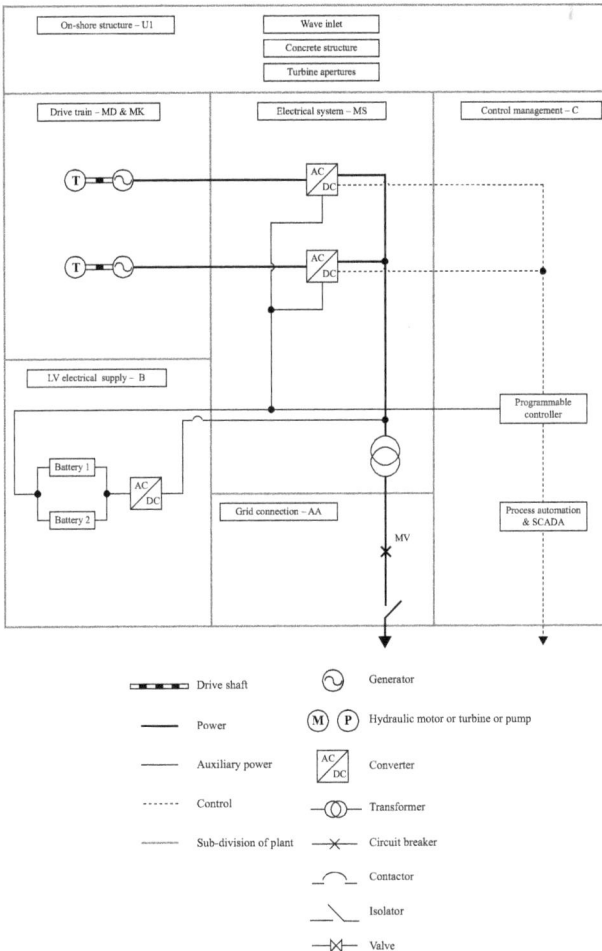

Figure C.1(a) WECa, FBD, an on-shore oscillating water column device based loosely upon the Alstom, Wavegen, Limpet concept

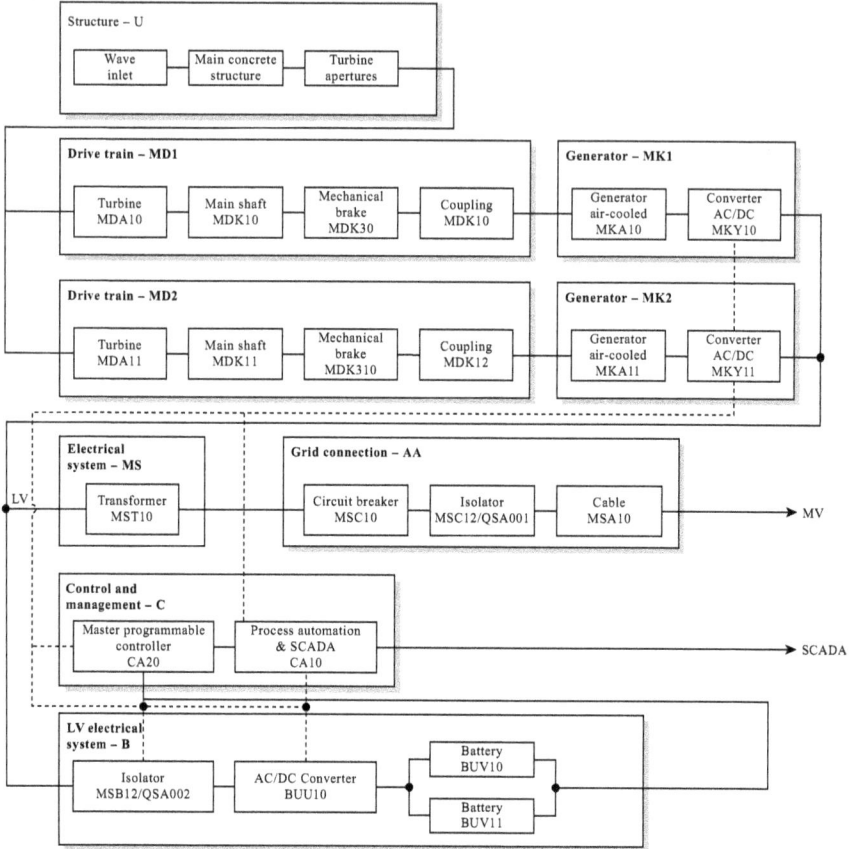

Figure C.1(b)   WECa, RBD

Table C.1   WECa, FMEA

System/Branch	Select 1 to %	VGB Code	Sub-assemblies	Qt	Data Source	λGi_min	λGi_max	Δλ FREcon100 %=λGi	Δλ FREcon=λGi_max	Env. adj. factor # zi	Adj. single (min) λGi_min*nEi	System Total @100% Power (min)	System Total @50% Power, Redundancy (min)	Adj. single (max) λGi_max*nEi	System Total @100% Power (max)	System Total @50% Power, Redundancy (max)
FIXED STRUCTURE (US)	1	UA10	Wave Inlet Structure	1	Dekom (2014)	0.0011	0.0011	0.001	0.001	1.0	0.001			0.001		
	2	UA20	Main Concrete Structure	1	Dekom (2014)	0.001	0.099	0.001	0.099	1.0	0.001	0.004	0.004	0.100	0.101	0.101
	3	UA31, UA32	Turbine Aperture Structure	3	Dekom (2014)	0.003	0.030	0.003	0.100	1.0	0.003			0.100		
CORROSION PROTECTION (AB)	4	AB	Connecting links	1	Dekom (2014)	0.100	0.185	0.100	0.185	1.7	0.170	0.170	0.170	0.315	0.315	0.315
DOUBLE DRIVE TRAIN (MDA-MK)	5	MDA10	Fixed pitch rotor blades for variable speed turbines for MEC application, per blade	2	Dekom (2014)	0.008	0.08	0.016	0.160	1.0	0.016			0.160		
	6	MDK10, MDK10, UP00, MDK40	Main shaft, main bearing, couplings	2	Dekom (2014)	0.031	0.055	0.062	0.110	1.0	0.062			0.110		
	7	MDK11	Shaft seals	2	Dekom (2014)	0.061	0.061	0.123	0.123	1.0	0.123			0.123		
	8	MDK20, MDV10	Gearbox, Lubrication & cooling system	2	Spinato, et al.(2009)	0.200	0.300	0.400	0.600	1.0	0.400	2.381	1.587	0.600	3.035	1.022
	9	MDK30, MDX10	Hydraulically Operated Brake System	2	Dekom (2014)	0.055	0.135	0.110	0.270	1.0	0.110			0.270		
	10	MKA10	High speed geared drive water-cooled wound or cage induction synchronous generator	2	Tavner (2012)	0.200	0.250	0.400	0.500	1.0	0.400			0.500		
	11	MKY10	LV Fully-rated VSC in the power range 2-3 MVA to convert generator outputs to fixed LV power frequencies	2	Chung, et al.(2015)	0.593	0.593	1.186	1.186	1.0	1.186			1.186		
	12	CA04, MKY11	VSC Controller	2	Chung, et al.(2015)	0.042	0.042	0.084	0.084	1.0	0.084			0.084		
LV ELECTRICAL SUPPLY (BU)	13	BUU11	LV Contactor	1	IEEE Gold Book (1999)	0.0052	0.0052	0.005	0.005	1.0	0.005			0.005		
	14	BUU10	LV Fully-rated DC/AC VSC	1	Chung, et al.(2015)	0.063	0.063	0.063	0.063	1.0	0.063			0.063		
	15	BUU12	LV Circuit Breaker feeding Auxiliary Supply	1	IEEE Gold Book (1999)	0.003	0.003	0.003	0.003	1.0	0.003	0.143	0.093	0.003	0.375	0.249
	16	BUU13	LV Cables for Auxiliary System	1	IEEE Gold Book (1999)	0.004	0.004	0.008	0.008	1.0	0.008			0.008		
	17	BUV10	Battery Electrical Auxiliary System for MEC	1	Dekom (2014)	0.032	0.147	0.064	0.294	1.0	0.064			0.294		
ANCILLARY SYSTEM (XA)	18	XAA10	Ventilation System	2	Dekom (2014)	0.105	0.105	0.210	0.210	1.0	0.210	0.397	0.364	0.310	0.597	0.364
	19	RAA30	Water-cooled Heat Exchanger	2	Dekom (2014)	0.094	0.094	0.187	0.187	1.0	0.187			0.187		
CONTROL & MANAGEMENT (CA)	20	CA03, MKA11	Device Controller & Output Gate Controller for MEC application	1	Dekom (2014)	0.239	0.630	0.239	0.630	1.0	0.239	0.525	0.525	0.630	1.384	1.384
	21	CA10	Process Automation & SCADA	1	Dekom (2014)	0.286	0.754	0.286	0.754	1.0	0.286			0.754		
GRID CONNECTION (AA)	22	CA10	SCADA monitoring interface, near-shore or shore-one for MEC application & Umbilical fibre optic cable	1	Dekom (2014)	0.010	0.016	0.010	0.016	1.0	0.010			0.016		
	23	AAG30	MV Circuit breaker feeding MEC output to power grid	1	IEEE Gold Book (1999)	0.006	0.006	0.006	0.006	1.0	0.006	0.019	0.019	0.006	0.029	0.029
	24	MDX30	Hydraulic Hose terminations on-shore	1	This book	0.001	0.005	0.001	0.005	1.0	0.001			0.005		
	25	AAG31	Dry-type MV/LV Transformer to step down output voltage for LV auxiliary system	1	IEEE Gold Book (1999)	0.002	0.002	0.002	0.002	1.0	0.002			0.002		
DEVICE TOTAL FAILURE RATE, λes138% (Failures/unit/year)				38		3.143	3.636	3.469	5.882		3.439	3.439	2.665	5.081	5.081	4.563
RELIABILITY SURVIVOR FUNCTION, R(1year), %						11.73%	3.44%	1.03%	0.39%		2.03%	1.03%	6.90%	0.34%	0.34%	1.27%

Simple Failure Rate Sum | Un-adjusted Failure Rate Estimates | Environmentally adjusted Minimum Failure Rate Estimate | Environmentally adjusted Maximum Failure Rate Estimate

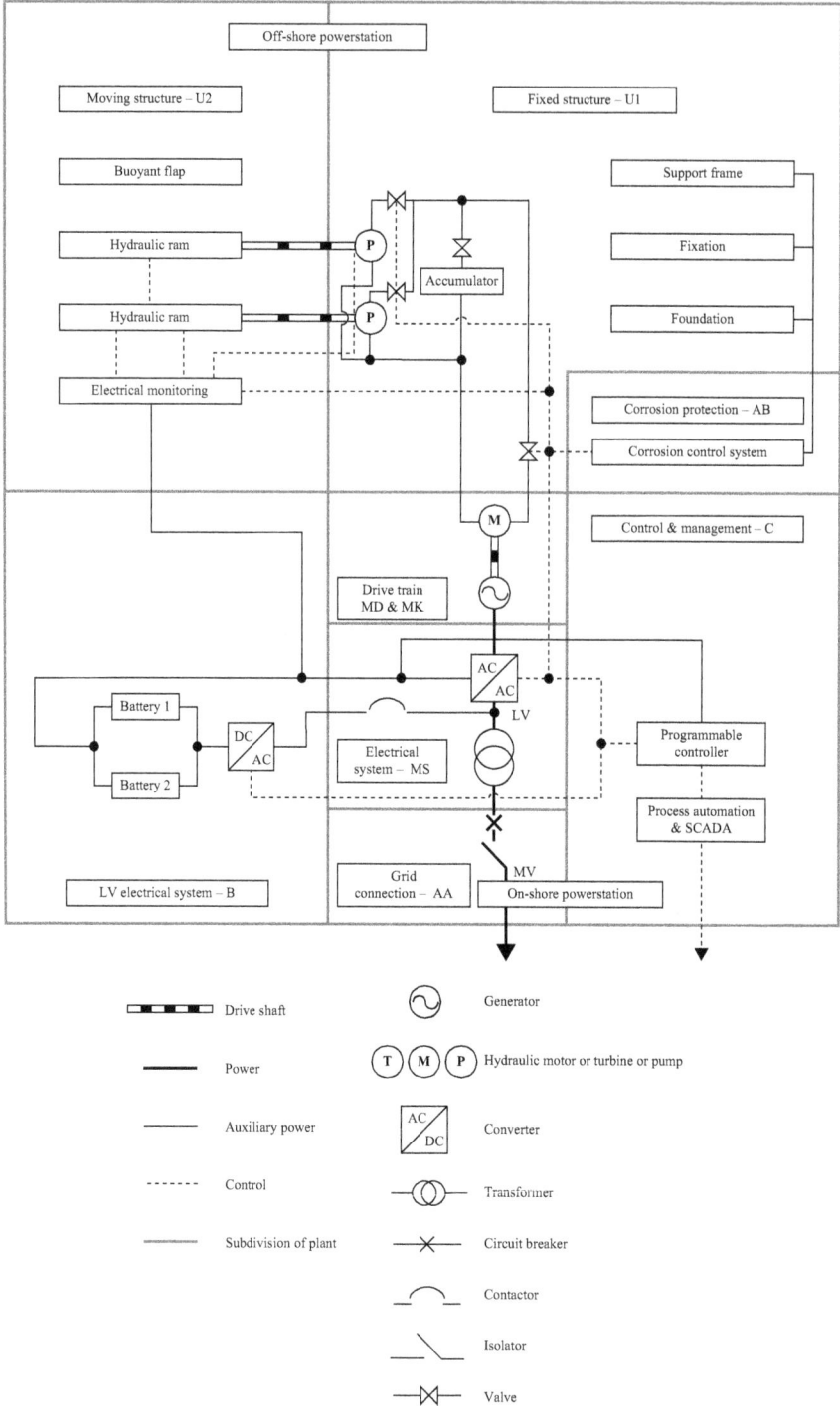

*Figure C.2(a)   WECb, FBD, A near shore Oscillating Wave Surge Converter
based loosely upon the Voith, Aquamarine, Oyster concept*

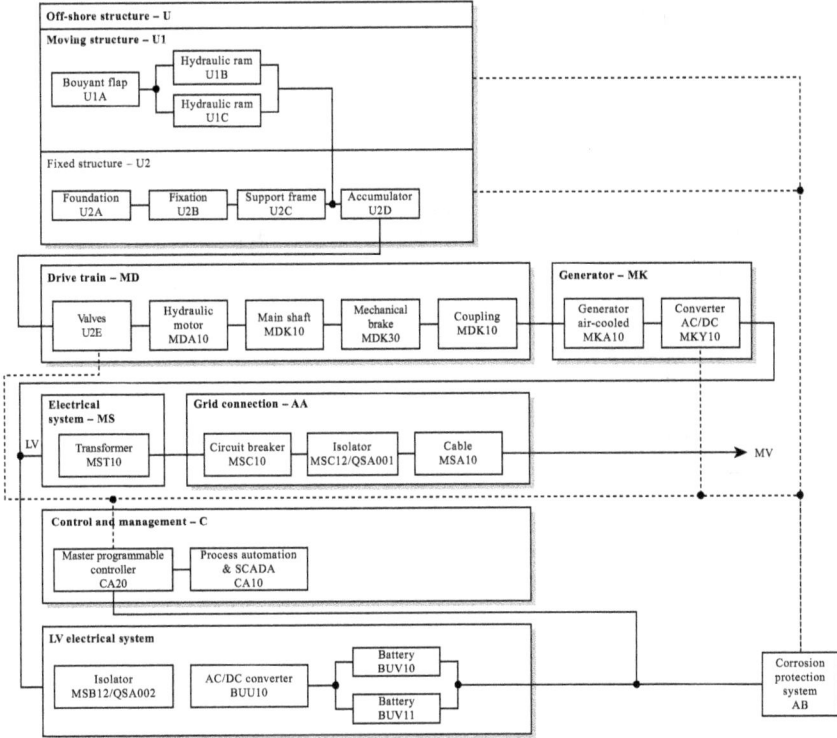

Figure C.2(b)   WECb, RBD

Table C.2   WECb, FMEA

RELIABILITY CHARACTERISTICS					Surrogate data failure rate (Failures/year)		Un-adjusted Failure Rate Estimates			Environmentally adjusted Minimum Failure Rate Estimates			Environmentally adjusted Maximum Failure Rate Estimates		
System/Branch	Extent of redundancy (=1 for N)	VGB Code	Sub-assemblies	Number of Sub-assemblies, Qt	Data Source	λGt		Predicted Total Failure Rates (Failures/year)		Environment adjustment factor	Adjusted single subassembly failure rates	System Total @ 100% Power	System Total @ 50% Power, Accounting for Redundancy	Adjusted single subassembly failure rates	System Total @ 100% Power	System Total @ 50% Power, Accounting for Redundancy
						λGt_min	λGt_max	λt_FRRmin = λGt_min	λt_FRRmax = λGt_max	ef_xt	λGt_min * nfi			λGt_max * nfi		
OFFSHORE SYSTEM — FIXED STRUCTURE (US)	1	U1	Monopile or fixed sea-bed location and placing systems for MEC application	1	Delorm (2014)	0.001	0.050	0.050	0.050	2.0	0.004	0.004	0.004	0.165	0.165	0.165
MOVING STRUCTURE (US)	2	UMD11	Mooring Yoke System	1	Delorm (2014)	0.050	0.060	0.050	0.060	3.3	0.165			0.165		
	3	UbdU50 CA601 UB001 UL001 EA001	Buoyant, hinged steel structures for MEC application	1	Torner (2014)	0.0023	0.0022	0.002	0.002	2.9	0.007	0.202	0.168	0.007	0.202	0.168
	4	U1B-R/U1C	Subsea Hydraulic Rams	2	Torner (2014)	0.005	0.005	0.009	0.009	3.3	0.030			0.030		
HYDRAULIC SYSTEM (HUS)	5	MHX10	Hydraulic receivers, accumulators & tanks for MEC application	2	Delorm (2014)	0.039	0.039	0.079	0.079	3.3	0.260			0.260		
	6	MHX20, UMD40001, UMD20002, UA010003, UMD35004	Hydraulic Hoses to shore, assume 500 m length	4	This book	0.02	0.1	0.080	0.405	3.3	0.264	0.555	0.277	1.320	1.752	0.866
	7	MHX30	Hydraulic Hose terminations off-shore	4	This book	0.002	0.01	0.008	0.040	3.3	0.026			0.132		
	8	MHX40	Hydraulic Hose terminations on-shore	4	This book	0.001	0.005	0.004	0.020	1.0	0.004			0.020		
CORROSION PROTECTION (AB)	9	AB	Corrosion Protection	1	Delorm (2014)	0.044	0.117	0.044	0.117	1.7	0.075	0.075	0.075	0.199	0.199	0.199
ON-SHORE POWER SYSTEM — POWER TAKE OFF DRIVE (B)	10	MHX1, MHX2, MHX3	Hydraulic power units for MEC application	3	Torner (2012)	0.027	0.08	0.080	0.240	1.0	0.080			0.240		
	11	MDA10, MDA20 UP001	Fixed pitch rotor blades for variable speed turbines for MEC application, per blade	2	Delorm (2014)	0.008	0.080	0.016	0.160	1.0	0.016			0.160		
	12	MDK20, MDV10	Gearbox, Lubrication & cooling system	2	Spinato, et. al. (2009)	0.200	0.300	0.400	0.600	1.0	0.400	1.322	1.102	0.600	1.616	1.605
	13	MKA10	High Speed geared drive water-cooled wound or cage induction synchronous generator	1	Torner (2012)	0.300	0.350	0.300	0.350	1.0	0.400			0.500		
	14	MKY11, MKY12, MKY13	LV Partially-rated VSC in the power range 0.6-0.8 MVA to convert generator outputs at fixed LV power frequencies	3	Cheng, et al.(2013)	0.100	0.150	0.390	0.300	1.0	0.300			0.300		
	15	CA34 MKY1	VSC Controller	1	Cheng, et al.(2013)	0.042	0.042	0.126	0.126	1.0	0.126			0.126		
LV ELECTRICAL SUPPLY (B)	16	BUU10	LV Fully-rated VSC in the power range 1-2 MVA to convert generator outputs to fixed LV power frequencies	1	Cheng, et al.(2013)	0.593	0.593	0.593	0.593	1.0	0.593			0.593		
	17	BUU11, AAG10	MV or LV Isolator	1	IEEE Gold Book (1999)	0.005	0.005	0.005	0.005	1.0	0.005			0.005		
	18	BUU10	LV Fully-rated AC/DC VSC for auxiliary system	1	Cheng, et al.(2013)	0.063	0.063	0.063	0.063	1.0	0.063	0.736	0.490	0.063	0.966	0.644
	19	BUU12	LV Circuit Breaker feeding Auxiliary Supply	1	IEEE Gold Book (1999)	0.003	0.003	0.003	0.003	1.0	0.003			0.003		
	20	BUV10	Battery Mounted Auxiliary System for MEC	2	Delorm (2014)	0.032	0.147	0.064	0.294	1.0	0.064			0.294		
	21	BUU12	LV Cables for Auxiliary System	2	IEEE Gold Book (1999)	0.004	0.004	0.008	0.008	1.0	0.008			0.008		
ANCILLARY SYSTEM (XAA)	22	XAA30	Water-cooled Heat Exchanger	1	Delorm (2014)	0.094	0.094	0.281	0.281	1.0	0.281	0.595	0.397	0.281	0.595	0.397
	23	XAA10	Ventilation System	1	Delorm (2014)	0.105	0.105	0.314	0.314	1.0	0.314			0.314		
ELECTRICAL SYSTEM (ME)	24	AAG10	LV Fully-rated AC/DC VSC for auxiliary system	1	Cheng, et al.(2013)	0.063	0.063	0.063	0.063	1.0	0.063			0.063		
	25	MKY10	MV fully-rated single channel VSCs in the power range 5-7 MVA to convert generator outputs to fixed LV power frequencies	1	Carroll, et al.(2014)	0.740	0.740	0.740	0.740	1.0	0.740	0.821	0.821	0.740	0.893	0.893
	26	MDA10, MDA20	Fixed pitch rotor blades for variable speed turbines for MEC application, per blade	2	Delorm (2014)	0.008	0.080	0.008	0.080	1.0	0.008			0.080		
	27	CA10	MV or LV Isolator	1	IEEE Gold Book (1999)	0.005	0.005	0.005	0.005	1.0	0.005			0.005		
	28	CA11	LV Contactor	1	IEEE Gold Book (1999)	0.005	0.005	0.005	0.005	1.0	0.005			0.005		
CONTROL & MANAGEMENT (CA)	29	CA21, MKA11	Device Controller & Output Gain Controller for MEC application	1	Delorm (2014)	0.050	0.050	0.050	0.050	1.0	0.050	0.092	0.092	0.050	0.161	0.161
	30	CA10	Process Automation & SCADA	1	Delorm (2014)	0.042	0.111	0.042	0.111	1.0	0.042			0.111		
GRID CONNECTOR — GRID CONNECTION (AA)	31	UMD12	Cable Stationary Support with Swivel	1	Delorm (2014)	0.050	0.050	0.050	0.050	3.3	0.165			0.165		
	32	AAG01 XD3000	Sub-sea connector	1	Delorm (2014)	0.009	0.009	0.009	0.009	3.3	0.030			0.030		
	33	AAG01	Neptune LV or MV cable for MEC application, assume 100 m length	1	Delorm (2014)	0.042	0.111	0.042	0.111	3.3	0.139	0.361	0.361	0.367	0.638	0.638
	34	CA10	SCADA monitoring interface, sea-shore or shore-sea for MEC application & Umbilical fibre optic cable	1	Delorm (2014)	0.010	0.016	0.010	0.016	3.3	0.033			0.052		
	35	AAG01	Fixed LV export cable systems, assume 50 km length	1	IEEE Gold Book (1999)	0.024	0.024	0.024	0.024	1.0	0.024			0.024		
DEVICE TOTAL FAILURE RATE, λtotUPN (Failures/unit/year)	68					2.478	3.480	3.968	3.468		4.789	4.789	3.784	7.313	7.313	5.571
RELIABILITY SURVIVOR FUNCTION, R/year, N						6.27%	3.15%	1.89%	4.12%		0.86%	0.86%	3.27%	0.07%	0.07%	0.39%
						Simple Failure Rate Sum		Un-adjusted Failure Rate Estimates		Environment adjustment factor	Environmentally adjusted Minimum Failure Rate Estimates			Environmentally adjusted Maximum Failure Rate Estimates		

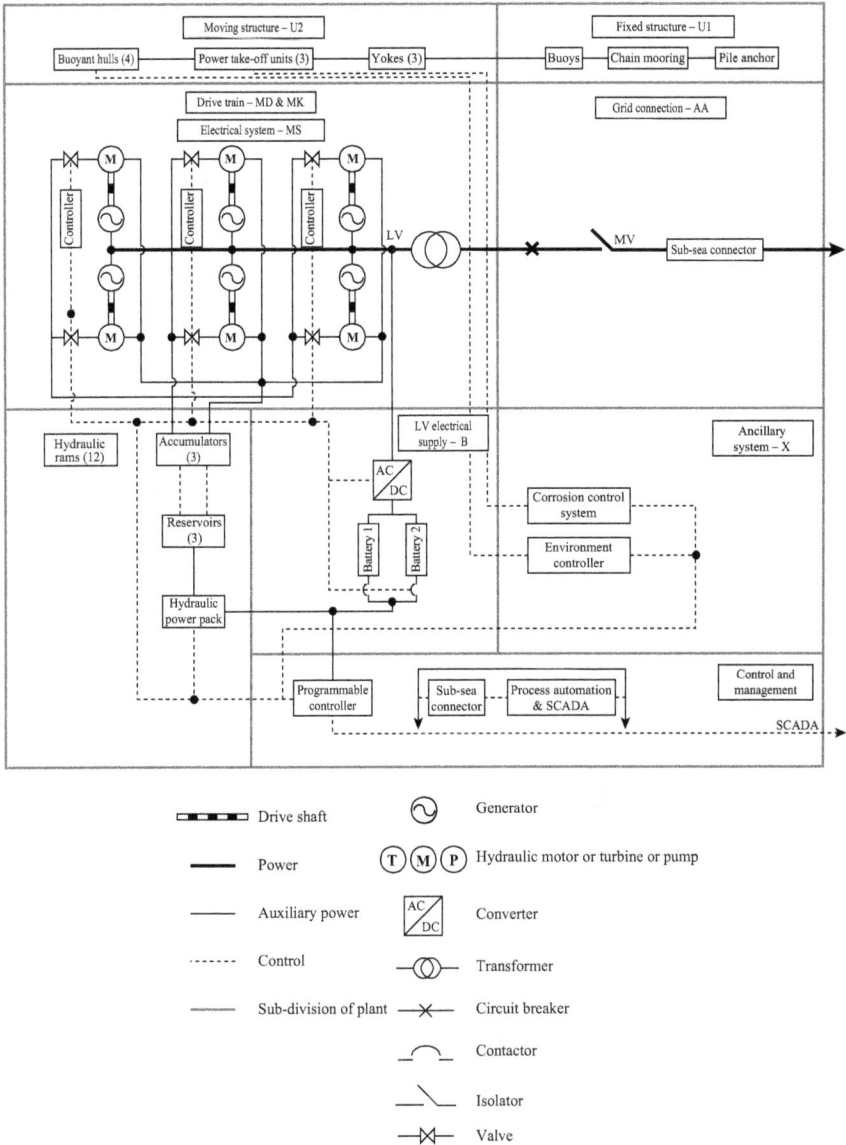

Figure C.3(a)   WECc, FBD, An off-shore Attenuating Wave Surge Absorber based loosely upon the Pelamis, concept

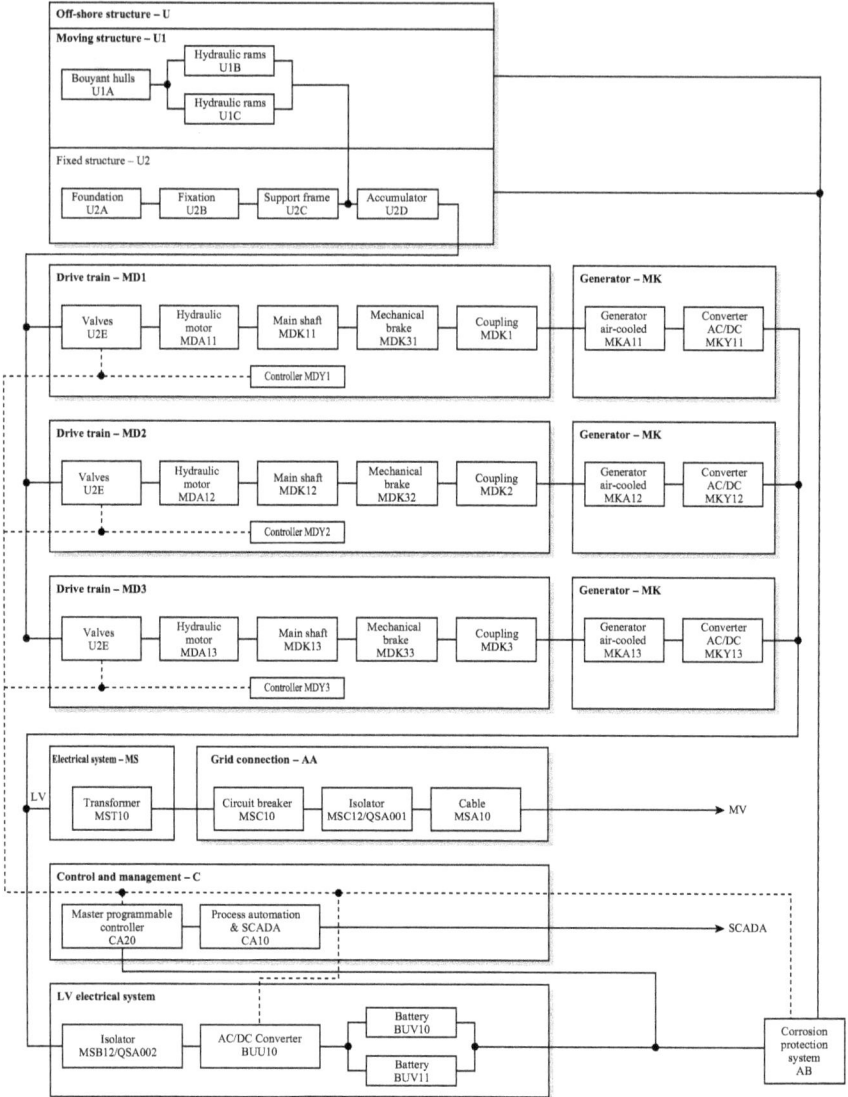

Figure C.3(b)   WECc, FBD

Table C.3   WECc, FMEA

RELIABILITY CHARACTERISTICS						Surrogate data failure rates (Failures/year)		Un-adjusted Failure Rate Estimates		Environmental adjustment factor	Environmentally adjusted Minimum Failure Rate Estimates			Environmentally adjusted Maximum Failure Rate Estimates		
System/Branch	Mon of unfavourability 1=1 dn/%	VGB Code	Sub-assemblies	Number of Sub-assemblies, Qt	Data Source	ΔGi	ΔGi	Predicted Total Failure Rates (Failures/year)		Environmental adjustment factor # st	Adjusted single subassembly rates	System Total @ 100% Power	System Total @ 50% Power, Accounting for Redundancy	Adjusted single subassembly rates	System Total @ 100% Power	System Total @ 50% Power, Accounting for Redundancy
						ΔGi_min	ΔGi_max	λi_FREcon = ΔGi_min	λi_FREcon = ΔGi_max			ΔGi_min * nEi			ΔGi_max * nEi	
OFFSHORE POWER STATION																
FIXED STRUCTURES (ST)		1	UMI1CIUA01	Connecting links	4	Delcam (2014)	0.102		0.400		3.3	1.330					
		2	UL001	Catenary mooring, chain	4	Delcam (2014)	0.024	0.185	0.095	0.740	3.3	0.315	1.790	1.154	2.442	2.442	1.628
		3	LP001	Catenary mooring, steel wire rope+	4	Delcam (2014)	0.008		0.090		3.3	0.000					
		5	UMD14	Pile anchors	4	Delcam (2014)	0.027		0.050		3.3	0.097					
MOVING STRUCTURE (CI)		6	UMD11	Mooring Yoke System	1	Delcam (2014)	0.050	0.050	0.050	0.050	3.3	0.165	0.447	0.372	0.165	0.447	0.372
		7	UMB30 UA001 UB001 UL001 BA001	Buoyant hull for MEC application	4	Delcam (2014)	0.001	0.001	0.004	0.004	3.3	0.015			0.015		
		8	U1A	Yoke for Hydraulic Rams	3	Tavner (2016)	0.009	0.009	0.027	0.027	3.3	0.089			0.089		
		9	U1B & U1C	Sub-sea Hydraulic Rams	13	Delcam (2014)	0.005	0.005	0.054	0.054	3.3	0.178			0.178		
HYDRAULIC SYSTEM (MIX)		10	MDX10	Hydraulic reservoirs, accumulators & tanks for MEC application	6	Delcam (2014)	0.059	0.059	0.237	0.237	1.0	0.237	0.355	0.177	0.237	0.355	0.177
		11	MDX30 LP001	Hydraulic power units for MEC application	3	Delcam (2014)	0.039	0.039	0.118	0.118	1.0	0.118			0.118		
CORROSION PROTECTION (AB)		12	AB	Corrosion Protection	1	Delcam (2014)	0.044	0.127	0.044	0.117	1.7	0.075	0.075	0.075	0.199	0.199	0.199
POWER TAKE OFF UNITS		13	MKK1, MKK2, MKK3	Hydraulic power units for MEC application	3	Tavner (2012)	0.027	0.06	0.160	0.480	1.0	0.160	1.912	0.677	0.480	2.832	0.944
		14	MKA11, MKA12, MKA13	High Speed geared drive water-cooled wound or cage induction synchronous generator	6	Tavner (2013)	0.200	0.3	1.200	1.800	1.0	1.200			1.800		
		15	MKY11, MKY12, MKY13	LV Partially-rated VSC in the power range 0.6-0.8 MVA to convert generator output to fixed LV power frequencies	3	Chung, et al.(2015)	0.100	0.100	0.300	0.590	1.0	0.300			0.590		
		16	MKY20	VSC Controller	3	Chung, et al.(2015)	0.042	0.042	0.126	0.126	1.0	0.126			0.126		
LV ELECTRICAL SUPPLY (B)		17	MKC10, QA001	LV Fully-rated AC/DC VSC for auxiliary system	2	Chung, et al.(2015)	0.063	0.063	0.126	0.126	1.0	0.126	0.143	0.095	0.126	0.273	0.249
		18	BUU11, AAG10	MV or LV Isolator	1	IEEE Gold Book (1999)	0.005	0.005	0.005	0.005	1.0	0.005			0.005		
		19	BUU10	LV Fully-rated AC/DC VSC for auxiliary system	1	Chung, et al.(2015)	0.063	0.063	0.063	0.063	1.0	0.063			0.063		
		20	BUU11	LV Circuit Breaker feeding Auxiliary Supply	1	IEEE Gold Book (1999)	0.003	0.003	0.003	0.003	1.0	0.003			0.003		
		21	BUV10	Battery Electrical Auxiliary System for MEC	2	Delcam (2014)	0.032	0.147	0.064	0.294	1.0	0.064			0.294		
		22	BU	LV Cables for Auxiliary System	2	IEEE Gold Book (1999)	0.004	0.004	0.008	0.008	1.0	0.008			0.008		
ANCILLARY SYSTEM (XA)		23	XAA20	Water-cooled Heat Exchanger	3	Delcam (2014)	0.084	0.084	0.281	0.281	1.0	0.281	0.595	0.397	0.281	0.595	0.397
		24	XAA10	Ventilation System	3	Delcam (2014)	0.105	0.105	0.314	0.314	1.0	0.314			0.314		
GRID CONNECTION		25	UMD12	Cable Buoyancy Support with Swivel	1	Delcam (2014)	0.050	0.050	0.050	0.050	3.3	0.165	0.361	0.361	0.165	0.638	0.638
		26	AAG11 XG001	Sub-sea connector	1	Delcam (2014)	0.009	0.009	0.009	0.009	3.3	0.030			0.030		
		27	AAG12	Buoyant LV or MV cable for MEC application, assume 100 m length	1	Delcam (2014)	0.042	0.111	0.042	0.111	3.3	0.139			0.367		
		28	CA10	SCADA monitoring interface, one-shore or shore-to-sea for MEC application & Umbilical fibre optic cable	1	Delcam (2014)	0.010	0.016	0.010	0.016	3.3	0.033			0.052		
		29	AAG10	Fixed LV export cable systems, assume 10 km length	1	IEEE Gold Book (1999)	0.024	0.024	0.024	0.024	1.0	0.024			0.024		
ELECTRICAL SYSTEM (MS)		30	AAG01	Dry type LV/MV Transformers to raise export cable voltage for power output to shore-side substation	1	IEEE Gold Book (1999)	0.004	0.004	0.004	0.004	1.0	0.004	0.015	0.015	0.004	0.015	0.015
		31	AAG02	MV Circuit breaker feeding the shore-side sub-station	1	IEEE Gold Book (1999)	0.006	0.006	0.006	0.006	1.0	0.006			0.006		
		32	MBC10 QA	MV or LV Isolator	1	IEEE Gold Book (1999)	0.005	0.005	0.005	0.005	1.0	0.005			0.005		
CONTROLS MANAGEMENT (CA)		33	CA10	Device Controller & Output Data Controller for MEC application	1	Delcam (2014)	0.239	0.630	0.239	0.630	1.0	0.239	0.525	0.525	0.630	1.384	1.384
		34	CA11	Process Automation & SCADA	1	Delcam (2014)	0.286	0.754	0.286	0.754	1.0	0.286			0.754		
DEVICE TOTAL FAILURE RATE, λmc100% (Failures/unit/year)				89		1.723	3.068	4.372	4.756		6.158	6.158	3.889	9.380	9.380	6.063
RELIABILITY SURVIVOR FUNCTION, R(t/year), N						17.88%	4.69%	1.26%	0.85%		0.21%	0.21%	2.32%	0.01%	0.01%	0.23%
						Simple Failure Rate Sum		Un-adjusted Failure Rate Estimates			Environmentally adjusted Minimum Failure Rate Estimates			Environmentally adjusted Maximum Failure Rate Estimates		

C.2   Tidal stream devices

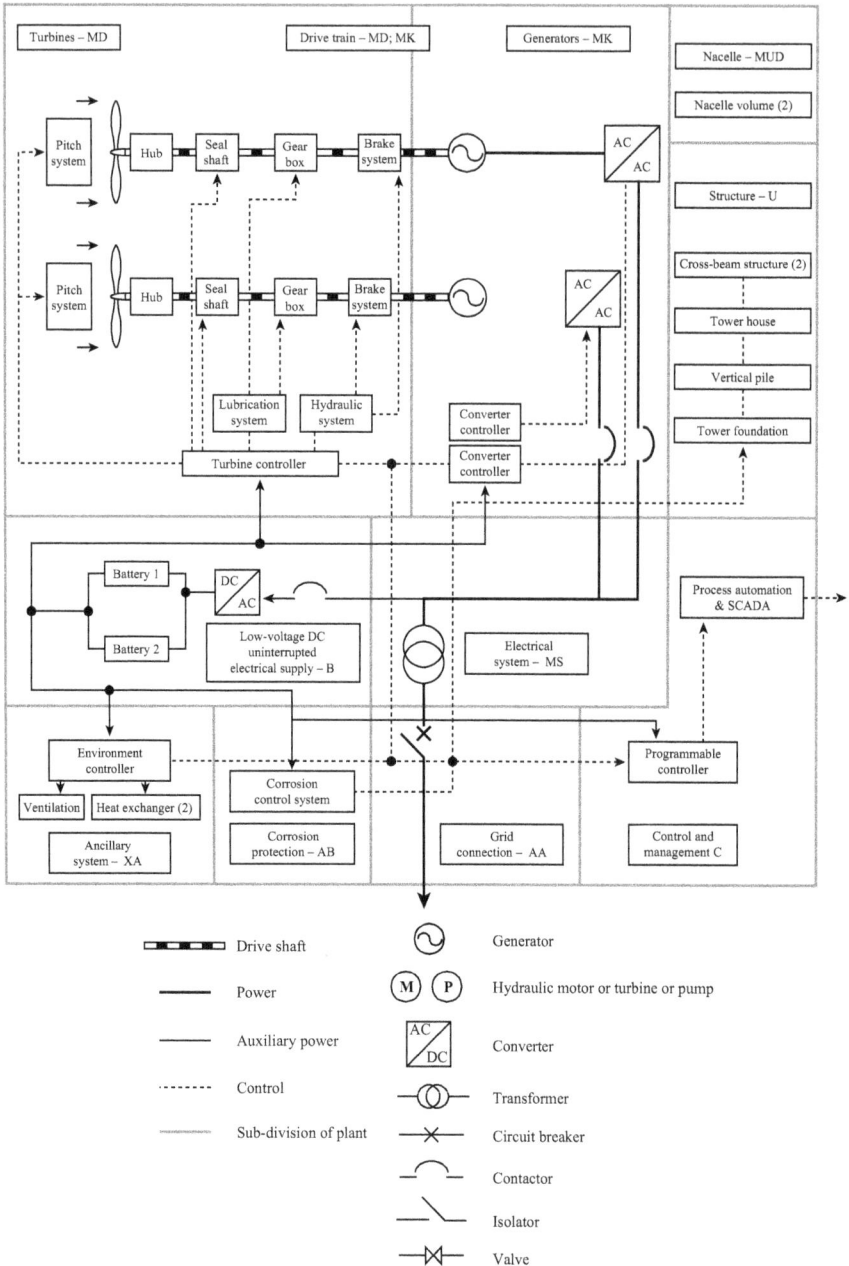

Figure C.4(a)   WECb, FBD, a near shore oscillating wave surge converter based loosely upon the Voith, Aquamarine, Oyster concept and (b) WECb, RBD

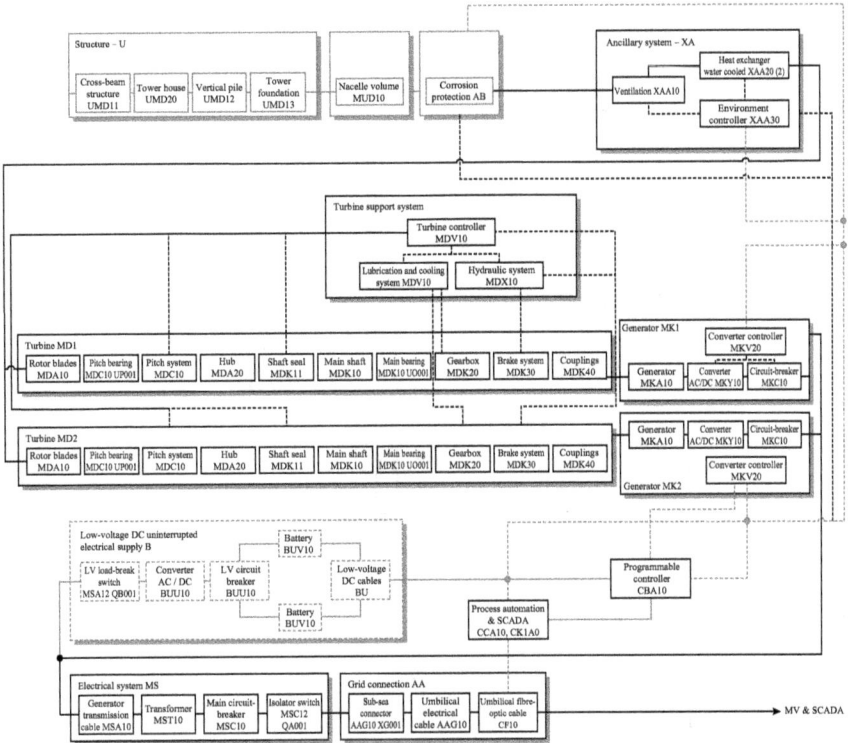

Figure C.4(b)   WECb, RBD

Table C.4   TSDa, FMEA

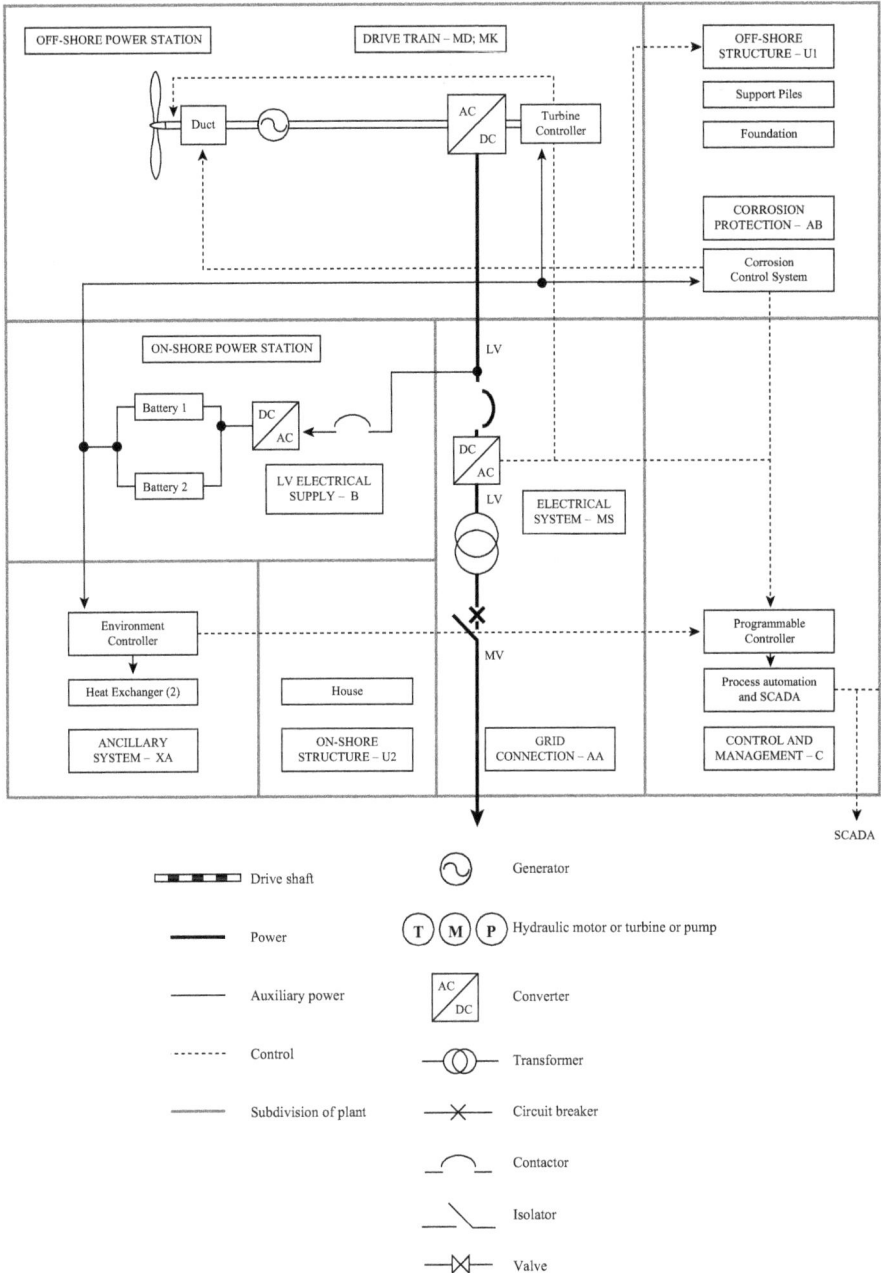

OFF-SHORE POWER STATION

DRIVE TRAIN – MD; MK

OFF-SHORE
STRUCTURE – U1

Support Piles

Foundation

Duct

AC
DC

Turbine
Controller

CORROSION
PROTECTION – AB

Corrosion
Control System

ON-SHORE POWER STATION

LV

Battery 1

DC
AC

DC
AC

LV

Battery 2

LV ELECTRICAL
SUPPLY – B

ELECTRICAL
SYSTEM – MS

MV

Environment
Controller

Programmable
Controller

Heat Exchanger (2)

House

Process automation
and SCADA

ANCILLARY
SYSTEM – XA

ON-SHORE
STRUCTURE – U2

GRID
CONNECTION – AA

CONTROL AND
MANAGEMENT – C

SCADA

Drive shaft

Generator

Power

T M P Hydraulic motor or turbine or pump

Auxiliary power

AC
DC Converter

Control

Transformer

Subdivision of plant

Circuit breaker

Contactor

Isolator

Valve

*Figure C.5(a)   TSDb, FBD, a fixed device incorporating a single, ducted,
horizontal axis, variable speed turbine based loosely upon the
Open Hydro concept*

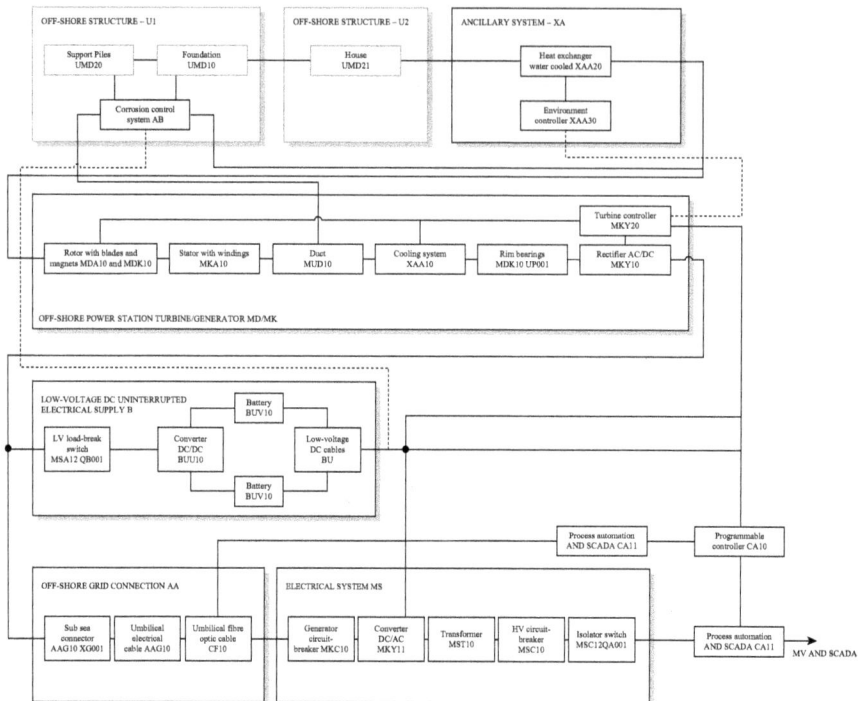

Figure C.5(b)   TSDb, RBD

Table C.5   TSDb, FMEA

RELIABILITY CHARACTERISTICS					Surrogate data failure rates (Failures/year)			Un-adjusted Failure Rate Estimate			Environmentally adjusted Minimum Failure Rate Estimate, No redundancy			Environmentally adjusted Maximum Failure Rate Estimate, No redundancy		
System/Branch	Ideal of redundancy 1=1 0=0 %	VGB Code	Sub-assemblies	Number of Sub-assemblies, Qi	Data Source	λGi_min	λGi_max	Predicted Total Failure Rate (Failures/year)		Environment adjustment factor #ei	Adjusted single ... rate	System Total @ 100% Power	System Total @ 100% Power	Adjusted single ... rate	System Total @ 100% Power	System Total @ 100% Power
								$\lambda i\_FREcon100\% = \lambda Gi$	$\lambda i\_FREcon = \lambda Gi\_max$	$\#ei$	$\lambda Gi\_min * nEi$			$\lambda Gi\_min * nEi$		
FIXED STRUCTURE (FS)	1	Ub0200, Ub8200	Civil engineering steel or concrete structures & foundations for MEC application	1	Deloux (2014)	0.0011	0.0011	0.0011	0.0011	3.3	0.004			0.004		
	2	UM0X1	Mantle or housing or duct structure	1	Deloux (2014)	0.0011	0.0011	0.0011	0.0011	3.3	0.004	0.009	0.007	0.004	0.007	0.007
CORROSION PROTECTION (AB)	3	AB10	Corrosion Protection	1	Deloux (2014)	0.044	0.117	0.044	0.117	1.7	0.075	0.075	0.075	0.199	0.199	0.199
DRIVE TRAIN (MDB/MK)	4	MDA10	Fixed pitch rotor blades for variable speed turbines for MEC application, per blade	10	Deloux (2014)	0.008	0.080	0.008	0.080	3.3	0.026			0.264		
	5	MDK10 UP601	Main bearing & seals	1	Deloux (2014)	0.021	0.021	0.021	0.021	3.3	0.069			0.069		
	6	MKA20	Direct drive water-cooled permanent magnet synchronous generator	1	Carroll, et al (2014)	0.071	0.071	0.071	0.071	1.7	0.120	0.219	0.219	0.120	0.458	0.458
	7	MUD10	Mantle or housing or duct structure	1	Deloux (2014)	0.001	0.001	0.001	0.001	3.3	0.004			0.004		
GRID CONNECTION (AAG)	8	AAG10	Fixed LV export cable systems, assume 10 km length	1	IEEE Gold Book (1999)	0.024	0.024	0.024	0.024	1.7	0.040	0.057	0.057	0.040	0.057	0.057
	9	CBA10	SCADA monitoring interface, sea-shore or shore-sea for MEC application & Umbilical fibre optic cable	1	Deloux (2014)	0.010	0.010	0.010	0.010	1.7	0.017			0.017		
ELECTRICAL SYSTEM (MS)	10	MKY10	LV Fully-rated VSC in the power range 2-3 MVA to convert generator output to fixed LV power frequency	1	Cheng, et al (2015)	0.593	0.593	0.593	0.593	1.0	0.593			0.593		
	11	MSC10	LV Contactor	1	IEEE Gold Book (1999)	0.005	0.005	0.005	0.005	1.0	0.005			0.005		
	12	MTT10	Dry type LV/MV Transformers to raise export cable voltage for power output to shore-side substation	1	IEEE Gold Book (1999)	0.004	0.004	0.004	0.004	1.0	0.004	0.613	0.613	0.004	0.613	0.613
	13	MSC10	MV Circuit breaker feeding the sea-shore export cable	1	IEEE Gold Book (1999)	0.006	0.006	0.006	0.006	1.0	0.006			0.006		
	14	MSC11 QA001	MV or LV Isolator	1	IEEE Gold Book (1999)	0.005	0.005	0.005	0.005	1.0	0.005			0.005		
LV ELECTRICAL SUPPLY (BU)	15	MSA12 QB001	MV or LV Isolator	1	IEEE Gold Book (1999)	0.005	0.005	0.005	0.005	1.0	0.005			0.005		
	16	BUU10	LV Fully-rated AC/DC VSC for auxiliary system	1	Cheng, et al (2015)	0.063	0.063	0.063	0.063	1.0	0.063			0.063		
	17	BUU11	LV Circuit Breaker feeding Auxiliary Supply	1	IEEE Gold Book (1999)	0.005	0.005	0.005	0.005	1.0	0.005	0.104	0.104	0.005	0.219	0.219
	18	BUV10	Battery Electrical Auxiliary System for MEC	1	Deloux (2014)	0.032	0.147	0.004	0.294	1.0	0.032			0.147		
	19	BU	Flexible LV transmission cable, assume 500 m length	1	IEEE Gold Book (1999)	0.001	0.001	0.001	0.001	1.0	0.001			0.001		
ANCILLARY SYSTEM (XA)	20	XAA10	Ventilation System	1	Deloux (2014)	0.105	0.105	0.105	0.105	1.0	0.105	0.198	0.198	0.105	0.198	0.198
	21	XAA30	Water-cooled Heat Exchanger	1	Deloux (2014)	0.094	0.094	0.094	0.094	1.0	0.094			0.094		
CONTROL & MANAGEMENT (EA)	22	CA10	Device Controller & Output Code Controller for MEC application	1	Deloux (2014)	0.239	0.630	0.239	0.630	1.0	0.239	0.525	0.525	0.630	1.384	1.384
	23	CA11	Process Automation & SCADA	1	Deloux (2014)	0.286	0.794	0.286	0.794	1.0	0.286			0.794		
DEVICE TOTAL FAILURE RATE , λtot100% (Failures/unit/year)				33		1.481	3.746	1.728	3.615		1.799	1.799	1.799	3.144	3.144	3.144
DEVICE RELIABILITY/SURVIVOR FUNCTION, R(t%), %						19.70%	6.42%	17.81%	3.70%		16.04%	16.04%	16.04%	4.31%	4.31%	4.31%
						Simple Failure Rate Sum			Un-adjusted Failure Rate Estimate		Environmentally adjusted Failure Rate Estimate	Environmentally adjusted Minimum Failure Rate Estimate, No redundancy			Environmentally adjusted Maximum Failure Rate Estimate, No redundancy		

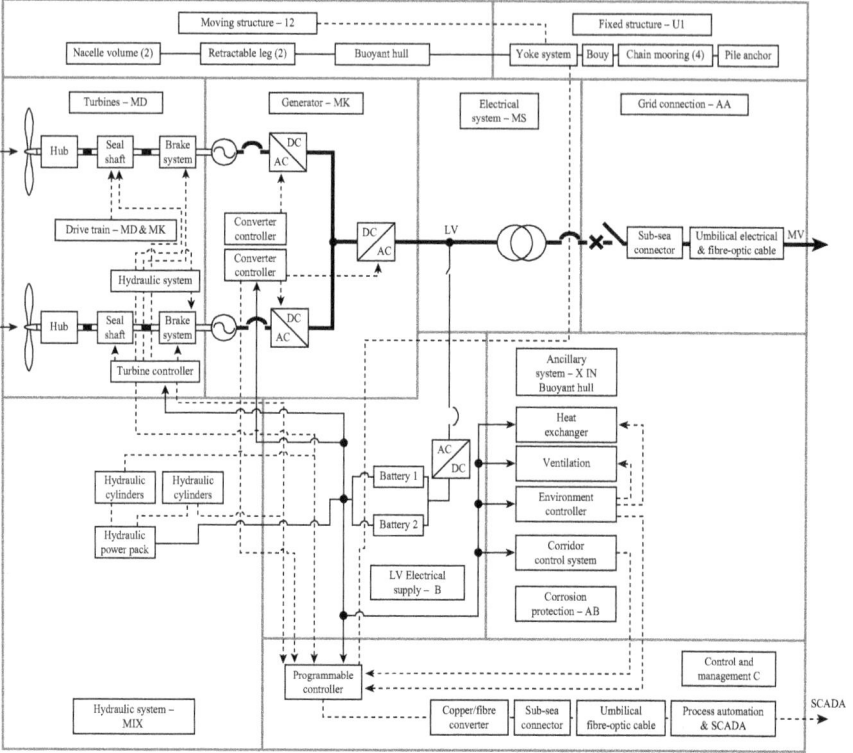

Figure C.6(a)   TSDc, FBD, A floating moored detachable device incorporating 2 horizontal-axis, fixed-speed, geared turbines based loosely upon the SRTT

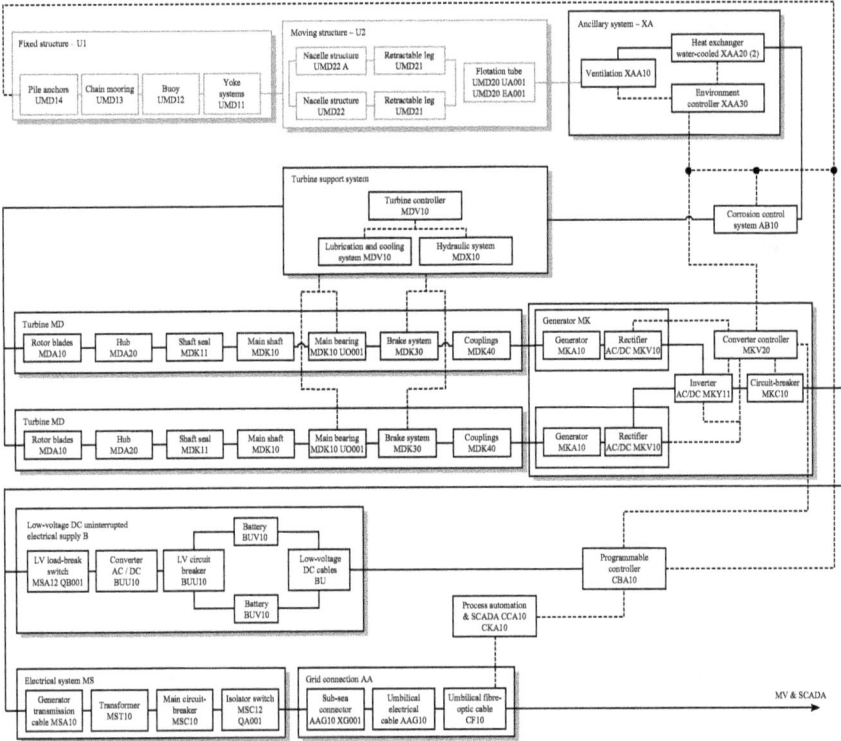

Figure C.6(b) TSDc, RBD

Table C.6 TSDc, FMEA

RELIABILITY CHARACTERISTICS					Surrogate data failure rates (Failures/year)		Un-adjusted Failure Rate Estimates		Environment adjustment factor	Environmentally adjusted Minimum Failure Rate Estimates			Environmentally adjusted Maximum Failure Rate Estimates		
System/Branch	Ident of subassembly 1–5 in N_s	VGB Code	Sub-assemblies	Number of Sub-assemblies, Qt	Data Source	λGt		Predicted Total Failure Rate (Failures/year)		Environment adjustment factor	Adjusted single subassembly failure rates	System Total @ 100% Power	System Total @ 50% Power, Accounting for Redundancy	Adjusted single subassembly failure rates	System Total @ 100% Power	System Total @ 50% Power, Accounting for Redundancy
						λGt_min	λGt_max	λL_FREicon10 0% = λGt	λL_FREicon = λGt_max	nEt	λGt_min * nEt			λGt_max * nEt		
FIXED STRUCTURE (SH)	1	UMD1 FIA010 1	Connecting Bolts	4	Delsers (2014)	0.190		0.466		4.0	1.820					
	2	UL001	Catenary mooring, chain	4	Delsers (2014)	0.024	0.115	0.095	0.540	5.3	0.511	1.710	1.154	2.443	2.443	1.638
	3	UP001	Catenary mooring, steel wire rope	4	Delsers (2014)	0.009		0.008		3.3	0.060					
	4	UMD14	Pile anchors	4	Delsers (2014)	0.007		0.035		3.3	0.097					
MOORING STRUCTURE (UD)	7	UMD11	Mooring Yoke System	1	Delsers (2014)	0.050	0.550	0.050	0.050	3.3	0.165			0.165		
	8	UMD20 UA001 UB001 UL001 RA001	Buoyant hull for MEC application	1	Delsers (2014)	0.001	0.001	0.001	0.001	3.3	0.004	0.255	0.157	0.004	0.255	0.157
	9	UMD21 UA001 UB001 UL001	Retractable Turbine Legs	2	Delsers (2014)	0.009	0.009	0.018	0.018	3.3	0.059			0.059		
	10	UMD22 UP001	Nacelle or housing or duct structure	2	Delsers (2014)	0.001	0.001	0.002	0.002	3.3	0.007			0.007		
HYDRAULIC SYSTEMS (HD)	11	MDX10	Hydraulic couplings, accumulators & tanks for MEC application	2	Delsers (2014)	0.019	0.030	0.079	0.079	1.0	0.079	0.118	0.079	0.079	0.118	0.079
	12	MDX20 GP001	Hydraulic power units for MEC application	1	Delsers (2014)	0.039	0.079	0.039	0.039	1.0	0.039			0.039		
CORROSION PROTECTION (AB)	13	AB	Corrosion Protection	1	Delsers (2014)	0.044	0.117	0.044	0.117	1.7	0.075	0.075	0.075	0.199	0.199	0.199
DOUBLE DRIVE TRAIN (MDR-MEC)	14	MDA10	Fixed pitch rotor blades for variable speed turbines for MEC application, per blade	4	Delsers (2014)	0.008	0.08	0.032	0.320	3.3	0.104			1.058		
	15	MDA20	Hub, Electric pitch control unit, Pitch bearing, per blade,	2	Chang, et al (2015)	0.190	0.198	0.190	0.380	3.3	1.267			1.267		
	16	MDE11	Shaft seals	1	Delsers (2014)	0.061	0.063	0.123	0.125	3.3	0.405			0.405		
	17	MDK10, MDR10 UP001, MDK40	Main shaft, main bearing, couplings	1	Delsers (2014)	0.031	0.055	0.062	0.110	1.0	0.062	5.800	2.460	0.110	4.838	3.239
	18	MDR30	Hydraulically Operated Brake System	2	Delsers (2014)	0.055	0.110	0.110	0.270	1.0	0.110			0.270		
	19	MKA10	High Speed geared drive water-cooled permanent magnet synchronous generator	2	Tavne (2012)	0.200	0.250	0.400	0.500	1.0	0.400			0.500		
	20	MKY10	LV Fully-rated VSC in the power range 2-3 MVA to convert generator outputs to fixed LV power frequencies	1	Chang, et al (2015)	0.593	0.593	1.186	1.186	1.0	1.186			1.186		
	22	MKY30	VSC Controller	1	Chang, et al (2015)	0.042	0.042	0.042	0.042	1.0	0.042			0.042		
	23	MKC10	LV Contactor	1	IEEE Gold Book (1999)	0.003	0.003	0.003	0.003	1.0	0.003			0.003		
LV ELECTRICAL SUPPLY (B)	24	BRA10 QA001	MV to LV Isolator	1	IEEE Gold Book (1999)	0.005	0.005	0.005	0.005	1.0	0.005			0.005		
	25	BUU10	LV Fully-rated AC/DC VSC for auxiliary system	1	Chang, et al (2015)	0.003	0.003	0.003	0.003	1.0	0.003	0.143	0.095	0.003	0.373	0.249
	26	BUU11	LV Circuit Breaker feeding Auxiliary Supply	1	IEEE Gold Book (1999)	0.003	0.003	0.003	0.003	1.0	0.003			0.003		
	27	BUV10	Battery Electrical Auxiliary System for MEC	2	Delsers (2014)	0.032	0.147	0.064	0.294	1.0	0.064			0.294		
	28	BU	LV Cabin for Auxiliary System	1	IEEE Gold Book (1999)	0.004	0.004	0.008	0.008	1.0	0.008			0.008		
ANCILLARY SYSTEM (BA)	29	XAA20	Water-circuit Heat Exchanger	2	Delsers (2014)	0.094	0.094	0.187	0.187	1.0	0.187	0.292	0.292	0.187	0.292	0.292
	30	XAA10	Ventilation System	1	Delsers (2014)	0.105	0.105	0.105	0.105	1.0	0.105			0.105		
GRID CONNECTION (AAA)	31	UMD12	Cable Buoyancy Support with Swivel	1	Delsers (2014)	0.050	0.050	0.050	0.050	3.3	0.165			0.165		
	32	AAG10 X001	Sub-sea connector	1	Delsers (2014)	0.060	0.009	0.009	0.009	3.3	0.030			0.030		
	33	AAG10	Buoyant LV or MV cable for MEC application, nominal 100 m length	1	Delsers (2014)	0.042	0.111	0.042	0.111	3.3	0.139	0.361	0.361	0.347	0.638	0.638
	34	CPA10	SCADA monitoring interface, sub-shore or state-use for MEC application & Umbilical fibre optic cable	1	Delsers (2014)	0.010	0.016	0.010	0.016	3.3	0.032			0.052		
	35	MSA10	Fixed LV export cable systems, nominal 10 km length	1	IEEE Gold Book (1999)	0.024	0.024	0.024	0.024	1.0	0.024			0.024		
ELECTRICAL SYSTEM (MS)	36	MST10	Dry type LV/MV Transformers to raise export cable voltage for power output in shore side	1	IEEE Gold Book (1999)	0.004	0.004	0.004	0.004	1.0	0.004			0.004		
	37	MSC10	MV Circuit breakers feeding the shore side sub-station	1	IEEE Gold Book (1999)	0.006	0.006	0.006	0.006	1.0	0.006	0.015	0.015	0.006	0.015	0.015
	38	MSC10 QA	MV to LV Isolator	1	IEEE Gold Book (1999)	0.005	0.005	0.005	0.005	1.0	0.005			0.005		
CONTROL & MANAGEMENT (CA)	39	CA10	Device Controller & Output Gain Controller for MEC application	1	Delsers (2014)	0.239	0.650	0.239	0.650	1.0	0.239	0.525	0.525	0.650	1.384	1.384
	42	CA11	Process Automation & SCADA	1	Delsers (2014)	0.286	0.734	0.286	0.734	1.0	0.286			0.734		
DEVICE TOTAL FAILURE RATE, incl 90% (Failures/year)				64				4.206	6.264		7.095	5.151		10.555	7.875	
RELIABILITY SURVIVOR FUNCTION, R(1year), %								1.49%	0.19%		0.08%	0.59%		0.00%	0.04%	
					Simple Failure Rate Sum		Un-adjusted Failure Rate Estimates		Environment adjustment factor	Environmentally adjusted Minimum Failure Rate Estimates			Environmentally adjusted Maximum Failure Rate Estimates			

MEC reliability data collection, based on wind industry experience

D.1 Introduction

D.1.1 Background

WT manufacturers, operators, maintainers and investors agree that it is essential for WTs to have a high reliability to achieve a high capacity factor and availability and thereby deliver electricity at a low cost of energy. An important factor in achieving those objectives is that WTs, when designed, should have the highest possible reliability. Currently, the European wind industry is achieving WT availabilities on-shore of 96%–97% and off-shore of 90%–95%. It would be desirable to raise these availabilities and design for reliability would contribute to that aim. The same applies to MECs and of course off-shore WTS are MECs.

An important requirement of design for reliability is to be able to measure, predict and analyse WT reliability using accurately defined mean time to failure (MTTF), mean time to repair (MTTR) and mean time between failures (MTBF) data for WTs. These standard terms are defined by international standards.

The definition of the terminology and taxonomy of wind turbines and the collection of reliability data and its interrelationship with WT design, defined by IEC 61400, need to be standardised. It is also clear that in order to increase WT reliability, more and higher quality reliability data is needed from the wind industry, within limits of commercial confidentiality.

This appendix is a proposal from the EU FP7 ReliaWind Consortium for the standardisation of:

- Taxonomy of the wind turbine.
- English terminology for the naming of components.
- Methods for collecting reliability data from wind turbines in the field.
- Method for reporting failures from wind turbines in the field.

The purpose of these standardisations is for the improvement of wind turbine reliability in the field, to raise wind turbine availability and lower the consequent cost of energy.

These issues also affect other industries, including off-shore oil and gas, power generation, transportation, military and aerospace.

An example of reliability data collected from the first of these industries, oil and gas, is shown by OREDA (1984–2015).

A standard for the collection of reliability data from that industry also exists, EN ISO 14224-2006.

D.1.2   Methods previously developed for the wind industry

The most detailed previous public domain WT data collection campaign was funded from 1996–2006 by the German Federal Ministry for Economics & Technology under the 250 MW Wind Test Programme, which included the Wissenschaftliche Mess- und Evaluierungsprogramm (WMEP), Scientific Measurement and Evaluation Programme, administered now by Fraunhofer IWES Institute.

This is built on earlier work by Schmid *et al.* (1991). A standard failure report form was used by WT operators for return to IWES.

Schmid *et al.* (1991) also gave valuable examples of data collection forms. The proposals below have drawn from this experience.

D.2   Standardising wind turbine taxonomy

D.2.1   Introduction

This section summarises the general principles and guidelines on which the taxonomy will be based and is derived from a deliverable prepared for the EU FP7 ReliaWind Consortium by the author and other consortium members.

The taxonomy should be adaptable for application to the common reliability analyses needed for WTs, such as failure mode and effects criticality analysis, failure rate Pareto analysis, reliability growth analysis and Weibull analysis.

The intention of adopting such a taxonomy would be to overcome current deficiencies of the data collection which can be summarised as follows:

- Consistency of naming of the systems, sub-systems, assemblies, sub-assemblies and components of WTs;
- Non-traceability of the system monitored;
- Unspecified WT technology or concept;
- Problems of confidentiality between parties when exchanging data.

D.2.2   Taxonomy guidelines

A WT taxonomy is a structure which names the main features of a WT in a standard terminology exemplified in Figure D.1.

- The taxonomy must be reliability-oriented, particularly in respect of analysis. It is agreed that such an approach is the best compromise between the various industry needs, which leads to a different system breakdown, grouping and terminology than would be achieved by simple components description.

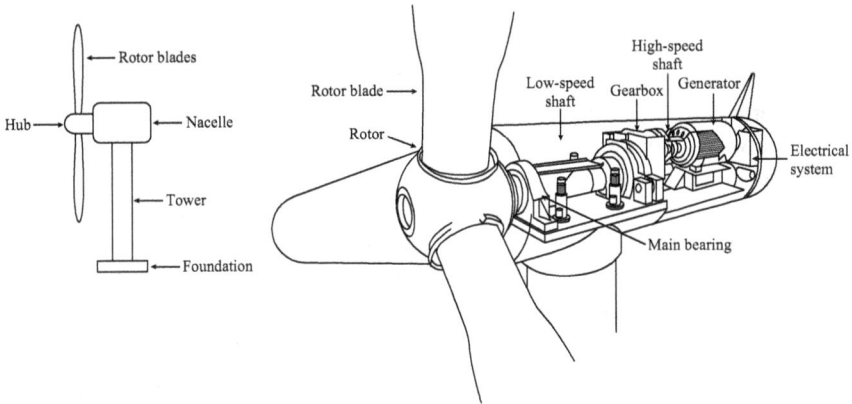

Figure D.1   Example of a WT and nacelle layout with its terminology

- The taxonomy will include all the WT concepts components in five levels. Data will be retrieved on the basis of a concept code, which allows WT model mapping within the taxonomy for any given data set.
- The taxonomy is based on a Danish Concept WT. That is an upwind, three-bladed, horizontal axis, un-ducted WT. Other concepts could be included upon achievement of significant industrial uptake by this taxonomy.
- At the highest level, outside the taxonomy, the WT concept should be identified by a code. For example indicating stall-, active stall- or pitch-regulated, fixed or variable speed, geared- or direct-drive, doubly-fed induction, induction or squirrel cage induction or wound or permanent-magnet synchronous generator. Therefore, each item in the taxonomy will be clearly linked to a code associated with each WT concept.
- The taxonomy should also inform the structure of the monitoring input/output (I/O) applied to the WT, whether that is for signal condition and data acquisition (SCADA) or condition monitoring system (CMS) signals and alarms because the taxonomy will be used to focus on SCADA and service log data available from operational wind farms. Therefore, the terminology of components in the input/output (I/O) list of the SCADA should agree with the component names used in the taxonomy above.
- The taxonomy shall be organised in five indented levels. Each level should be justified with a brief description that shall include the rationale for the level grouping and intended use.
- The first five indented levels of the taxonomy must comply with the table given in Section D.2. The taxonomy may not reach the lower level components, for example to individual electronic capacitors, but an analyst could add additional lower levels, if needed, but they must be compatible with the upper five levels of the taxonomy. Analysts could also add additional elements in levels 1–5, if absolutely necessary for their purpose, denoted by the prefix CUSTOM, although this customisation is strongly discouraged.

- The taxonomy will have a short code alpha-numeric designation for each item at each level. It is anticipated that the construction of the designation could follow the guidance of VGB PowerTech (2007), which adopts an alpha-numeric code although some in the wind industry prefer a word code.
- The lowest level components will be grouped according to the following two concepts:
  - Functional grouping for the signalling, supervisory and control components, examples: pitch encoder grouped with control and communication system, LV electrical systems grouped together.
  - Positional grouping for mechanical components, examples: gearbox, pitch system, blade, frequency converter, generator, blade.

For example, generator temperature sensor and pitch encoder are both components of the monitoring system. This segregation is necessary due to the nature of WT systems that signalling, supervisory and control components tend to spread throughout the WT, whereas mechanical devices are located in a specific position within the WT. This is exemplified in Table D.1.

- In case of ambiguity, the designation will follow the order above: first the functional groupings, second the positional grouping.
- At the lowest indented level, the component name should have no ambiguity with similar components of different assemblies, for example, the pitch pinions and the yaw pinions.

D.2.3   Taxonomy structure

The structure of system, sub-system, assembly, sub-assembly and component that should be adopted is shown in Figure D.2. The WT itself is considered as the system.

Examples of this terminology are shown in Table D.2.

Table D.1   Examples of parts groupings

Functional grouping	Positional grouping
Control and communication system	Generator
Lightning protection system	Pitch system
110 V Electrical auxiliary system	Gearbox
220 V Electrical auxiliary system	Yaw system
400 V Electrical auxiliary system	Blade
WT power system	Hub
SCADA system	Main shaft set
Collection system	Foundation
Grid connection	Tower
Hydraulics system	

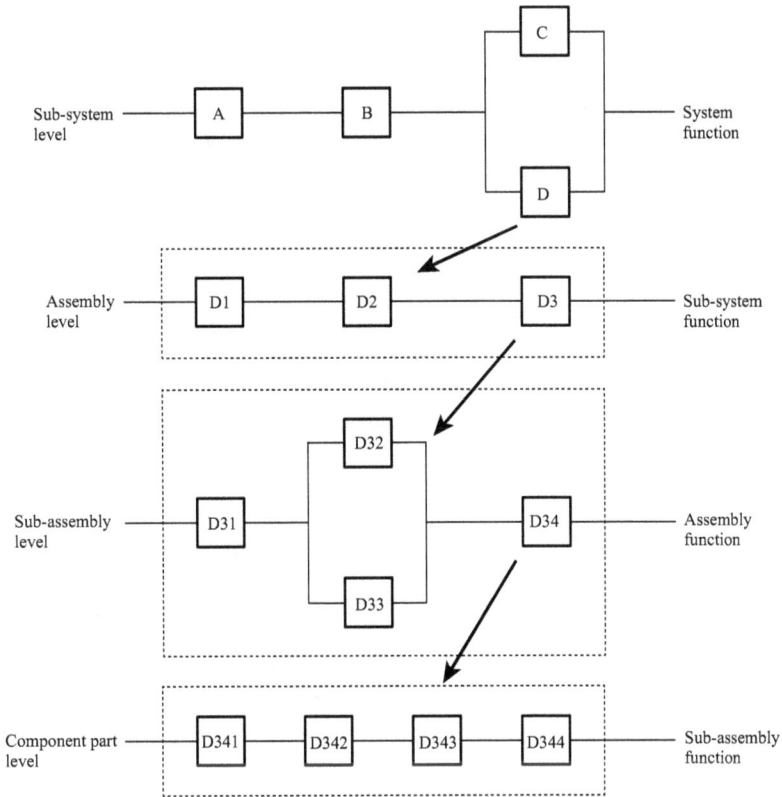

Figure D.2   Example of system, sub-system, assembly, sub-assembly and component structure, cf. Figure 2.7

Table D.2   Example of application of terminology

System	Sub-system	Assembly	Sub-assembly	Component
Wind turbine	Rotor	Electrical pitch system	Pitch motor	Brush
Wind turbine	Drive train module	Gearbox	Gearbox	Stage 1 Planetary Wheel
Wind turbine	Electrical module	Frequency converter	Power electronics	IGBT

D.3   Standardising methods for collecting WT reliability data

In the ReliaWind Consortium, the following method was used where it was proposed that reliability data from WTs should be collected in five tables as follows:

Table D.3 Fields are defined as follows:

Wind farm	*Confidentiality will require that this be an anonymous identifier*
Turbine ID	*Turbine identifier within the wind farm*
Time of event	*Time stamp in ISO form yyyy-mm-dd hh:mm:ss*
Mean downtime (MDT)	*Total number of hours during which the turbine was not operational, i.e. includes all the time needed to restore the WT to an operating condition*
Time to repair (TTR)	*Actual number of hours completing the repair, i.e. excludes logistics associated with the repair action such as having the component delivered to site, arranging technicians' time*
System structure	*The system structure used is that set out in Section D.2*
Sub-system	*It should usually be possible to ascribe a failure to a particular sub-system*
Assembly	*It should usually be possible to ascribe a failure to a particular assembly*
Sub-assembly	*It may not always be possible to ascribe a failure to a particular component*
Part	*Select from an approved list. It may not always be possible to ascribe a failure to a particular component*
Failure mode	*The particular way in the failure occurred, independent of the reason for failure. This may be subjective, but very useful if available*
Root cause	*The cause of failure. This may be subjective, but very useful if available*
Maintenance category	*A description of the maintenance impact of the failure:* *1. Manual restart* *2. Minor repair* *3. Major repair* *4. Major replacement*
Severity category	*A description of the severity of the failure based on MIL-STD-1629A Section 4.4.3 which relates to the ability of the system to carry out the function for which it was defined safely and efficiently* *1. Minor* *2. Marginal* *3. Critical* *4. Catastrophic*
Additional information	*Pertinent comments where available*

This list of events in Table D.3 will be exhaustive within the following criteria:

- The event required manual intervention to restart the machine.
- The event resulted in downtime $\geq$ 1 hour.
- There will be no missing events or missing time periods; or if there are, the missing time periods will be noted and the reasons stated.

Table D.3 *Events*

Wind farm	Turbine ID	Date and time of event	Time to repair TTR (hours)	Actual repair time ART (hours)	System	Sub-system	Assembly	Sub-assembly	Component	Failure mode	Root cause	Maintenance category	Severity Category	Additional information
A	1	2008-04-01 11:28:01	54.2	N/A	Wind Turbine	Drive train	Gearbox assembly	Gearbox	N/A	N/A	N/A	4	3	N/A
A	23	2008-04-24 01:56:11	168.4	3.5	Wind Turbine	Rotor	Pitch system	N/A	N/A	N/A	N/A	3	2	N/A
B	2	2008-04-25 08:43:24	2.5	1	Wind Turbine	Power	Generator assembly	Generator	Stator phase b winding	Open	Over current	1	1	Series failure
...	...	...	...	...	...	...	...	...	...	...	...	...	...	...

• Every cell in the table should have either a data value or be filled with N/A, not available.

Table D.4 would be derived entirely from Table D.3 and no new information is added.

It is thought unlikely that enough detail will be available in Table D.3 to permit the calculation of failure rate on a per-component basis.

Failure rates should be reported per year as standard but information from Table D.3 which allows calculation per operational period in a year, per GWh in a year, per revolution, or some other metric, depending on what information is available for the particular wind farm.

Confidentiality may require that this information be aggregated on a wind farm, rather than WT basis.

This information could be presented graphically, for example as shown in Figure 3.6(a).

Table D.5 is also derived entirely from Table D.4 and no new information is added.

The downtime will be given in units of hours.

Confidentiality may require that this information be aggregated on a wind farm, rather than WT basis.

This information could be presented graphically, for example as shown in Figure 3.6(b).

Table D.4   Failure rates

Wind farm	Turbine	Sub-system	Assembly	Year			
				1	2	3	...
A	1	Drive Train	Gearbox	0	2	1	...
A	1	Power	Generator	2	1	2	...
A	1	Rotor	Pitch	1	2	1	...
...	...		...	...	...	...	...

Table D.5   Downtime

Wind farm	Turbine	Sub-system	Assembly	Year			
				1	2	3	...
A	1	Drive train	Gearbox	24	5	1	...
A	1	Power	Generator	65	4	2	...
A	1	Rotor	Pitch	21	5	5	...
...	...		...	...	...	...	...

There should be two types of Table D.5, Tables D.5(a) and D.5(b):

- Table D.5(a) should have all values stated exactly but must remain private to the operator, to maintain end user confidentiality; and
- Table D.5(b) should be available more widely to operational consortia but be less specific about machine characteristics, with identifiable parameters categorised in appropriate ranges to anonymize the data, as shown in the table above. Valuable component failure rate data is available to operators in Appendix E.

The control type column will be populated from a standard list.

Further columns may be added to this table depending on what information is available for each wind farm (Table D.6).

Confidentiality requirements may mean that the information in Table D.6 could not be publicly available.

For a wind farm to be included in the survey, it is desirable that the site contains at least 15 turbines that have been running for at least two years since commissioning.

Data for the tables above should be provided by WT operators.

D.3.1 Standardising downtime event recording

The approach recommended is to describe and classify downtime events as stoppages of duration ≥ 1 hour and requiring at least a manual restart, categorising downtime events as follows:

- Category 1: manual restart
- Category 2: minor repair
- Category 3: major repair
- Category 4: major replacement

Table D.7 gives another simple but valuable measurement for maintenance staff of cumulative turbine rotations, giving an insight into collective turbine operation and possible wear.

Table D.6 Wind farm configuration

Wind farm	Turbines	Rated power (MW)	Mean wind Speed (m.s⁻¹)	Mean turbulence intensity	Hub height (m)	Rotor diameter (m)	Terrain type	Control type	...
A	20–40	1–2	6–8	0.25–0.50	60	40	Off-shore	A	...
B	0–20	2–3	8–10	0.50–0.75	55	30	On-shore exposed	B	...
...	...	...	...	...	...	...	...	...	...

Table D.7 Additional turbine information

Wind farm	Month	Energy generated (GWh)	Revolutions	...
A	2008-01	50	1.544×10^5	...
A	2008-02	70	2.422×10^5	...
...	...	...	...	...

D.4   Standardising failure event recording

D.4.1   Failure terminology

When a failure has occurred, it is important to record the details of that failure. In the WMEP Failure Report Form shown in Appendix E, a simple tick box approach was adopted.

D.4.2   Failure recording

The intention of this section is to provide a broad method of failure recording, rather than trying to capture every different possible failure mode. For example, a bearing failure could encompass:

- Inner race failures;
- Outer race failures;
- Cage failures;
- Element failures.

The recommended failure recording terminology is in part recursive, referring successively to the component, sub-assembly, assembly, sub-system, system, defined in Section D.2, in turn.

D.4.3   Failure location

Location indicators are needed for components, such as bearings, where several may be found in a single assembly or sub-assembly, in that case, the following rules could be followed, using the failure example above:

- If more than one epicyclic stage exists in a gearbox, the first stage is that closest to the WT rotor and so on.
- In a parallel shaft gear train, the pinion drives and the gear is driven.
- The two ends of a gearbox are the rotor end or generator end.
- Where there are two bearings on a gearbox shaft, that closer to the gear should be referred to as the inner bearing, and that further from the gear as the outer bearing.
- Generator bearings should be defined as drive end (DE) and non-drive end (DE).

Appendix E

Reliability of key sub-assemblies, portfolio of surrogate data

Table E.1 summarises reliabilities for key MEC systems described above taken from the PSD sources set out in tables in Chapter 5, many of the sources above being collated in Delorm (2014), and these values are used in the following MEC reliability analyses in the book.

Table E.1   Reliabilities of key MEC systems

Sub-assemblies	Sources	Reported failure rate, failures/yr, Ground benign, GB, environment	
		Low	High
Civil engineering steel or concrete structures and foundations for MEC application	Delorm (2014)	0.0011	0.0500
Monopile or fixed sea-bed location and pinning systems for MEC application	Delorm (2014)	0.0011	0.0500
Nacelle or housing or duct structure	Delorm (2014)	0.0011	0.0011
Cross-beam structure	Delorm (2014)	0.0011	0.0500
Detachable sea-bed location and pinning systems for MEC application	Tavner (2016)	0.0022	0.0022
Buoyant, hinged steel structures for MEC application	Tavner (2016)	0.0022	0.0022
Buoyant hull for MEC application	Delorm (2014)	0.0011	0.0011
Sub-sea hydraulic rams	Tavner (2016)	0.0090	0.0090
Yoke for hydraulic rams	Tavner (2016)	0.0045	0.0045
Retractable turbine legs	Delorm (2014)	0.0090	0.0090

(Continues)

Table E.1   (Continued)

Sub-assemblies	Sources	Reported failure rate, failures/yr, Ground benign, GB, environment	
		Low	High
Mooring yoke system	Delorm (2014)	0.0500	0.0500
Connecting links	Delorm (2014)	0.1000	0.1850
Catenary mooring, chain	Delorm (2014)	0.0237	
Catenary mooring, steel wire rope	Delorm (2014)	0.0000	
Pile anchors	Delorm (2014)	0.0074	
Corrosion protection	Delorm (2014)	0.0443	0.1168
Ventilation system	Delorm (2014)	0.1048	0.1048
Water-cooled heat exchanger	Delorm (2014)	0.0936	0.0936
Hydraulic power units for MEC application	Delorm (2014)	0.0394	0.0394
Hydraulic receivers, accumulators and tanks for MEC application	Delorm (2014)	0.0394	0.0394
Hub, Electric pitch control unit, pitch bearing, per blade	Chung *et al.* (2015)	0.1950	0.1950
Hub, hydraulic pitch control unit, pitch bearing, per blade	Tavner (2012)	0.1000	0.2000
Main shaft, main bearing, couplings	Delorm (2014)	0.0310	0.0550
Rim bearing and seals	Delorm (2014)	0.0208	0.0208
Shaft seals	Delorm (2014)	0.0613	0.0613
Fixed pitch rotor blades for variable speed turbines for MEC application, per blade	Delorm (2014)	0.0080	0.0800
Pitchable rotor blades for variable or fixed speed turbines for MEC application, per blade	Tavner (2012)	0.0267	0.0800
Gearbox, lubrication and cooling system	Spinato *et al.* (2009)	0.2000	0.3000
Hydraulically operated brake system	Delorm (2014)	0.0550	0.1350
Direct drive water-cooled permanent magnet synchronous generator	Carroll *et al.* (2014)	0.0706	0.0706
High speed geared drive water-cooled wound induction synchronous generator	Tavner (2012)	0.2000	0.2500

(Continues)

Table E.1 (*Continued*)

Sub-assemblies	Sources	Reported failure rate, failures/yr, Ground benign, GB, environment	
		Low	High
High speed geared drive water-cooled permanent magnet synchronous generator	Tavner (2012)	0.2000	0.2500
LV partially rated VSC in the power range 0.6–0.8 MVA to convert generator outputs to fixed LV power frequencies	Chung *et al.* (2015)	0.1000	0.1000
LV fully rated VSC in the power range 2–3 MVA to convert generator outputs to fixed LV power frequencies	Chung *et al.* (2015)	0.5930	0.5930
Flexible LV transmission cable, assume 500 m length	IEEE Gold Book (1999)	0.0012	0.0012
MV or LV isolator	IEEE Gold Book (1999)	0.0052	0.0052
LV contactor	IEEE Gold Book (1999)	0.0052	0.0052
LV fully rated AC/DC VSC for auxiliary system	Chung *et al.* (2015)	0.0630	0.0630
LV circuit breaker feeding auxiliary supply	IEEE Gold Book (1999)	0.0027	0.0027
LV cables for auxiliary system	IEEE Gold Book (1999)	0.0040	0.0040
Battery electrical auxiliary system for MEC	Delorm (2014)	0.0319	0.1472
MV fully rated single-channel VSCs in the power range 3–5 MVA to convert generator outputs to fixed LV power frequencies	Carroll *et al.* (2014)	0.7400	0.7400
MV fully rated parallel redundant channel VSCs in the power range 4–6 MVA to convert generator outputs to fixed LV power frequencies	Carroll *et al.* (2014)	0.4000	0.4000
MV contactor	IEEE Gold Book (1999)	0.0153	0.0153
VSC controller	Chung *et al.* (2015)	0.0420	0.0420
Device controller and output gain controller for MEC application	Delorm (2014)	0.2390	0.6300
Process automation and SCADA	Delorm (2014)	0.2860	0.7540

(Continues)

Table E.1   (Continued)

Sub-assemblies	Sources	Reported failure rate, failures/yr, Ground benign, GB, environment	
		Low	High
Dry type LV/MV transformers to raise export cable voltage for power output to shore-side substation	IEEE Gold Book (1999)	0.0036	0.0036
Dry type MV/LV transformer to step down output voltage for LV auxiliary system	IEEE Gold Book (1999)	0.0018	0.0018
Oil-cooled MV/HV transformers to raise export cable voltage for power output to the grid	IEEE Gold Book (1999)	0.0030	0.0030
LV circuit breaker feeding the sea-shore export cable	IEEE Gold Book (1999)	0.0044	0.0044
MV circuit breakers feeding the shore-side sub-station	IEEE Gold Book (1999)	0.0064	0.0064
SCADA monitoring interface, sea-shore or shore-sea for MEC application and umbilical fibre optic cable	Delorm (2014)	0.0101	0.0159
Cable buoyancy support with swivel	Delorm (2014)	0.0500	0.0500
Buoyant LV or MV cable for MEC application, assume 100 m length	Delorm (2014)	0.0422	0.1112
Sub-sea connector	Delorm (2014)	0.0000	0.0091
Fixed LV export cable systems, assume 10 km length	IEEE Gold Book (1999)	0.0236	0.0236
Fixed MV export cable systems, assume 10 km length	IEEE Gold Book (1999)	0.0052	0.0052
Shore-side sub-station	IEEE Gold Book (1999)	0.0200	0.0200
MV circuit breaker feeding the sea-shore export cable	IEEE Gold Book (1999)	0.0064	0.0064
MV circuit breaker feeding MEC output to power grid	IEEE Gold Book (1999)	0.0064	0.0064
Sub-station power flow controller	Chung *et al.* (2015)	0.0420	0.0420

A clean energy technology timeline

Years BCE/AD up to the Common Era	Origin, location, originators	Technologies including: • Wind devices; • Wave devices; • Tidal devices; • Machines; • Control; • Electronics; • Reliability	Photo, diagram or note
		Early history	
200 BCE	Persia, vertical axis wind machines first used.		
70 AD	Heron, Greece, Pneumatica, Development of a model reaction steam turbine. There is debate whether Heron invented this model, or whether it was stimulated from different examples developed by Heron? The model is a first-century turbine. By the end of the nineteenth century, this device led to steam and then gas turbine power, first for electrical power stations, then for ships and aircraft.	
634–644 AD	Persia, Iran Practical windmills built in Sistan, Persia-Afghanistan border region by the Rashidun Caliph Umar for grinding grain and pumping water. In 1963, 50 of them were still operating in Neh, Iran. Hau (2006)	Vertical axis wind turbines (VAWT), efficiency <10%, long vertical axle driven shafts, rectangular blades. Enclosed by a two-storey circular wall, millstones at the bottom, rotor at the top. Rotor: spoked with 6–12 upright ribs, each covered with cloth to form separate sails. Photo, 1900	

(Continued)

c650–900	Arrival in Scotland, England, Wales and Ireland of people from Northern Europe	Sutton Hoo, Anglo-Saxon ship burial.	
		Typical Danish Viking ship used to invade England and Ireland.	
1090	China, Kaifeng, Henan Province (河南开封) Development by Su Sung (苏松) of the world's first water-driven mechanical clock, incorporating an escapement mechanism.	E rise of technology in China, An exemplar water-driven mechanical clock but dismantled and not repeated. Published with permission from the Needham Research Institute	
1119	Netherlands	Four-blade horizontal axis wind turbine (HAWT) post-mills for milling grain, draining water and sawing wood. Easy to yaw, but support might have been an issue. Post-mills dominated milling and pumping in Europe until the nineteenth century.	
1219	China	Reference to a possible common origin for windmills in Western China or Persia, see above, Needham (1994) and Zhang (2009). Sheng Ruozi refers to wind-mills in the 'Placid Retired Scholar', Yelü Chucai (耶律初) 1190–1244, prominent Jin and Yuan statesman after fall to the Mongols in 1234. The passage refers to a journey in 1219 to Turkestan, modern Xinjiang (新疆) and of Hechong Fu (和沖) to Samarkand, modern Uzbekistan. These VAWT wind-mills had luffable self-adjusting sails, in response to wind conditions as the wind-mill rotated.	

(Continued)

	cf. earlier reported Chinese turbines.	Yancheng City, Jiangsu province (盐城市, 江苏). Recent reconstruction of traditional Chinese lightweight wood VAWT, for water pumping, fitted with vertically luffable sails, to adjust power, in the maritime tradition.
		Recent reconstruction of a small traditional Chinese lightweight wood HAWT, for water pumping. Again, luffable sails.
1200s	Europe cf., earlier 1119 Netherlands cap-mill.	Four-blade, squat HAWT structure, with wooden shutters adjustable as luffable blades.
1295	Netherlands	Four-blade HAWT upwind cap-mill. Advantage of cap-mill over post-mill is that it is unnecessary to turn the whole buck or mill body into the wind. This allows more space in the tower for machinery and storage.
1424	China	Admiral Zheng He (鄭和) sailed this eight-mast ship from China to Palembang in Sumatra. His ship is here compared to Columbus' three-mast and bowsprit *SANTA MARIA,* 1492.

(*Continued*)

1700	UK	Relatively recent horizontal axis watermill for milling flour.	
1750–1780	Abraham Darby, UK Beginnings of the European Industrial Revolution	UK's earliest established an Iron Works at Coalbrookdale, Staffordshire for processing iron, wrought iron and steel. Including manufacture of the components for the Ironbridge at nearby Coalbrookdale, designed by Darby.	
1776	James Watt, UK	Double acting stationary steam engine for water pumping and draining mines. Watt's design built by D. Napier & Son, London, 1832, Superior Technical School of Industrial Engineers, Madrid, Spain.	

(Continued)

| 1829 | Robert Stephenson and Co, UK | *Rocket*, development of successful coal-fired locomotive for Rainhill trials on the Manchester to Liverpool railway. Designed by Robert Stephenson, with advice from George Stephenson and Henry Booth, manufactured in Newcastle-upon-Tyne. |  |

Coal, oil and gas pollution sources combine

Nineteenth-century industrial activities in the UK

Industrialisation and coal burning, Black Country

Railways

Lot's Road Power Station, London

Steelworks, Sheffield

(Continued)

		Start of electrical age	
1820–1873	Oersted, Faraday, Arago, Babbage, Herschel, Pixii, Lenz, Wheatstone, Cooke, Gramme, Pacinotti, Europe and USA	Development of the first DC electric generators and motors using current switching via rotating commutators.	
1882	Edison, USA	Beginning of the electric age. First central power station developed with large DC generators and underground LVDC cable for large-scale transmission of DC electric power from Pearl Street Power Station, central New York. Efficiency <4%.	
1885	Ferranti, UK, Subsequently, Tesla & Westinghouse, USA	Beginning of the electric age. First central power station with AC generators, transformers and underground HVAC cables, from Deptford Power Station to Central London, for large-scale transmission and usage of AC electric power. Efficiency <10%.	

(Continued)

		Cleaner energy the push-back starts	
1878	Armstrong, UK	Sir William Armstrong, solicitor, engineer and industrial magnate, Newcastle-upon-Tyne with a passion for water-power, as opposed to use of coal. Armstrong installed a hydro-electric plant, at 'Cragside' his Northumberland home, using lake water and a 4.5 kW water-turbine-driven DC dynamo to light lamps in the house. Worked with local colleagues William Siemens and Joseph Swan.	
1886	Tesla, USA, then Dolivo-Dobrovolsky, Russia, Germany and Switzerland	First polyphase AC induction motor/generator developed. Three-phase AC induction motor developed, working independently from Tesla. Then, developing a three-phase AC induction motor with a polyphase slip-ring rotor connection to resistors for starting control.	
1887	Charles Parsons, Newcastle, UK.	4 kW, steam turbine driving a 2-pole DC generator with commutator.	
	Prof James Blyth, UK, Scotland, Royal College of Science, Glasgow, for electricity production and battery charging.	10 kW, 7-blade VAWT, driving a DC dynamo. Believed to have had adjustable blades, which contemporary alternatives did not. This is a large turbine, note Mrs Blyth standing below the VAWT.	

(Continued)

1891	Denmark, Poul la Cour developed a wind turbine for electricity production and battery charging.	10 kW, 4-blade HAWT, fixed pitch wind turbine driving a DC dynamo. Innovative aerodynamic system and electric utilisation for electrolysis and hydrogen storage, mitigating wind variability with 20 year life and high reliability. This is a large turbine, note man standing below the HAWT.	
1904–1911	USA and Germany, Ward Leonard, Kramer, Schrage	Development of variable speed DC and then AC electric generators/motors using switching via rotating commutators.	
1910–1914	Competition against electric windmills from internal combustion engines, following the development of diesel- and then petrol-engine driven generators and low cost of oil.		
1914–1918	First World War, reduction in oil supplies, 20–35 kW electric windmills being built.		
1918 post-war	Wind-mill develop-ment languished	Small wind turbines proving less reliable for electricity production than diesel- or petrol-engine driven generators, but AC electrical grid connection becoming more widespread.	
1920s	Germany, Betz	Following the First World War, influence of aerodynamic knowledge from aircraft starts to affect wind turbine design due to wing and propeller developments.	
1923–1973	Various inventors and industrial companies	Development of valve (ignitron) and semiconductor (thyristor) switches to convert AC/DC and DC/AC, facilitating reversible power flow through AC generators.	
1930s	Denmark, USA, Electric windmills become common on large farms.	Cheap high-tensile steel, windmills on prefabricated open steel lattice towers	Decline of American multi-blade turbine concepts in favour of European 4, 3, 2 and even 1-blade concepts.
1930s–1940s	Europe and North America, national AC electrification using fossil-fired power stations feeding national grids, based on Ferranti and Tesla's ideas with Westinghouse's investments. Research programmes in Denmark, France, Germany and UK considered wind power to supplement the national grid.		

(Continued)

1940	Germany, Ventimotor company formed by Ulrich Hutter. Cf., earlier light weight Chinese turbines.	W34 2-blade downwind HAWT cap-mill. Lightweight, cost-effective, guyed tower and typical excellent German aerodynamic design. Test centre near Weimar developing wind turbines for the war effort. Hutter (1954). This is a large turbine, note the man climbing the HAWT.	
1900	UK, Isle of Man Cf., 1943, China large Norias.	Laxey Hydraulic Wheel, water-driven steel structure, driving pumps to draw water from a lead, copper, silver and zinc mine. Note two people above the water wheel.	
1943	China, Gansu Province (甘肃). Cf., 1993, 3 Gorges Hydro Water Turbine, 700 MW, 50 years later!	Historic 1 kW vertical axis hydraulic turbine, for seed-milling or pumping, efficiency <5%. Published with permission from the Needham Research Institute	
	China, Chengdu, Szechuan Province (四川成都).	Historic traditional river-powered, large, three wooden, high-lift water Norias, 1–2 kW. For irrigation, efficiency <5%. Published with permission from the Needham Research Institute	

(*Continued*)

	Cf.,1900, Laxey mine water pump. and 2021, 55 kW diesel-driven agricultural irrigation pump.	Two historic traditional river-powered, large quantity, wooden, high-lift water Norias, 5 kW. For irrigating large fields. Published with permission from the Needham Research Institute	
1943	China, Nei Chiang, Szechuan Province (四川内江). Cf, 55 kW diesel-driven agricultural irrigation pump, 2021.	An historic traditional low output wooden water raising pump. Pedal-driven, human-powered square-pallet, chain clack conveyor for irrigating small rice paddy fields. Published with permission from the Needham Research Institute	
1944	China, Showing Chinese Industrial Revolution, following European and US with similar technologies.	Smelting iron in a cupola furnace at the National Resources Commission Machine Works. Lanzhou, Gansu (兰州 甘肃). Published with permission from the Needham Research Institute	
	Cf Abraham Darby's furnace, UK, 1790.	Steel production in an underground factory built into the mountain at the Central Machine Works. Kunming, Yunnan (昆明 云南). Published with permission from the Needham Research Institute	

(Continued)

1947	USA, Smith-Putnam, world's first megawatt-sizewind turbine connected to electrical distribution system Grandpa's Knob, Castleton, Vermont. Considered the grandfather of modern electric wind turbines.	1.25 MW, D 57 m, 40 m high, 2-blade downwind HAWT, geared, constant speed, full-span pitch control, stall-regulated, synchronous machine. Sophisticated, modern, AC grid-connected wind turbine. Design by P C Putnam [8], including aerodynamicist von Karman, MIT. Manufactured by S M Smith.	
1950–1970	Various inventors and industrial companies.	Development of new types of power semi-conductor switches, the thyristor, transistor, IGBT.	
1957 onwards	Number of inventors	Back-to-back reversible DC and AC drives using thyristor or other semiconductor bridges with forced static commutation, switched twice per cycle. Static commutator eventually renders rotating commutators obsolete, developments essential for variable speed electric wind turbines using AC generators and converters and with electric pitch motors.	
1966	France, La Rance tidal barrage.	First tidal barrage with a capacity of 240 MW and a capacity factor of 24%.	
1972	International oil crisis, triggered by Yom Kippur war, a renaissance of wind power, see Gipe (2016) considering both small and large wind turbines.		
1975	USA	General Electric invention of new fast-acting semiconductor devices	

(Continued)

1979		200 kW, D 24 m, three-blade upwind HAWT, geared-drive, fixed speed, stall-regulated with aerodynamic rotor blade tip brakes.	
	Denmark, Nibe, two experimental machines erected, incarnation of the modern wind turbine. One with pitch control and one without. If Smith-Putnam was the father of the modern electric wind turbine, these two were his strongest children.		
		630 kW, D 40 m, three-blade upwind HAWT, geared-drive, full-span pitch control, fixed speed, stall-regulated.	
1980s	Greece, Island of Crete.	Current example of historical HAWTs in rural Europe.	
	Cf., Earlier Chinese wind turbines	Traditional HAWT for water pumping on Lassithi plateau, with luffable sails.	

(Continued)

	Germany, Bottrop, on the Rhine-Herne Canal	Waste Water Treatment Plant (WWTP) pumping station using three parallel, electrically driven, Archimedes Screw water lift pumps, serving 1,350,000 people.	
1980s	USA, Great California Wind Rush, following oil price rises.	Large numbers of wind turbines $\leq$ 100 kW, mostly HAWT but some VAWT. Very poor reliability from many different designs but some did survive.	
1993	China, Hubei Province (湖北).	Three Gorges Dam Power Station, SanXiaDaBa (三峡大坝). Includes 32 × 700 MW, vertical axis hydraulic turbines driving	
	Cf., Gansu 1 kW Water Turbine above, 1943.	synchronous AC generators, installed within the dam. Efficiency 95%.	
	Cf, France-Spain IGBT HVDC Link 2022.	HVDC substation using thyristors feeding 1,200 MW, 500 kV DC from Three Gorges Dam GeZhouBa (葛洲坝) to ShangHai (上海) Converter station efficiency 90%.	

(Continued)

2003	UK	Marine current turbines Sea flow, 300 kW. ● Tested off the Devon Coast near Lynmouth; ● Subsequently withdrawn from service.	
2008	UK, Northern Ireland	Marine current turbines SeaGen twin tidal turbines; ● 1.2 MW; ● Tested in Strangford Loch, Northern Ireland; ● Subsequently withdrawn from service.	
	China, DongHai DaQiao (东海大桥), ShangHai (上海) Off-shore wind project. Turbines mounted on concrete gravity caissons. Note the low top-head mass and slender steel tower.	Installation of 3 MW off-shore three-blade upwind HAWTs, pitch-regulated, Sinovel SL3000/90, geared drive, with IGBT variable speed converters, grid-connected off-shore wind project.	

(Continued)

2014	Germany, BARD Off-shore Wind Project	400 MW off-shore wind farm, 80 5 MW three-blade, upwind HAWT; • Geared drive; • Doubly fed wound asynchronous generator; • Partially rated, grid-connected, IGBT variable speed converter; • Mounted on three-leg steel jackets, for deep water, rather than, more common earlier steel mono-piles. Wind turbines with step-up transformers connected by 36 kV AC collector cables to an off-shore sub- and IGBT HVDC converter-station. This inter-connects to shore from the HVDC converter-stations by 90 km, 170 kV DC cable links, enabling economic long-distance power transfer to a shore-based sub-station, where it is converted to AC and stepped-up to 380 kV.	
2014	UK, Orkney, EMEC test site.	Pelamis P2, 750 kW • Wave device under test; • Subsequently withdrawn from service.	

(Continued)

2015	UK, Cragside, Northumberland	Armstrong's *Cragside* house now includes a modern, water-powered, Archimedes screw driving at the top an 18.5 kW AC induction generator, supplying electricity to the house, replacing original 1870 DC electric installation, shown further above. Sir William Armstrong founded Armstrong College, subsequently King's College Durham University, becoming Newcastle University in 1964.	
2018	Canada, Bay of Fundy test site.	Open Hydro Tidal turbine; • Raised after being under test; • Subsequently withdrawn from service.	
2019	USA, Pacific Northwest National Laboratory (PNNL) supporting US Department of Energy's (DOE) Water Power Technologies Office	Tethys, Wave point absorber; • Wave device; • Prospective capacity, a few kW; • Future larger devices planned.	

(Continued)

2020	UK, Scotland. Green Ocean Energy	Wave Treader, prospective capacity 500–700 kW. • A wave energy device; • Designed to increase renewable energy yield from off-shore wind farms by accessing wave energy at turbine bases.	
2021	UK, Scotland, Orbital awarded a 30 MW Option Agreement from Crown Estate-Scotland for tidal energy project in Westray Firth, adjacent to EMEC, where O2 turbine is currently in operation.	Orbital Marine Power's O2, floating twin tidal turbine, capacity 2 MW. With lessons learnt from prior experience of fixed or floating devices, e.g.: • SeaFlow 300 kW fixed tidal turbine (2003); • SeaGen 1.2 MW fixed tidal device (2008); • Atlantis AK1000 • 1 MW fixed tidal turbine (2011); • Pelamis 750 kW floating wave device (2014).	
2021	UK, Scotland	Mocean Energy Blue X's, 20-metre long, 38-tonne floating prototype wave energy converter. • Peak capacity of 5–30 kW; • Successful five-month test period at sea off Orkney.	

(Continued)

2024	UK, Ireland, France, Germany and Spain,14-collaborators, coordinated by Ocean Energy in Ireland.	Wedusea Wave Energy Converter, prototype floating surface OE Buoy. 
		• Lower part open to sea, trapping air, wave pressure in submerged opening, excites water oscillation;
		• This drives trapped air through an air turbine generating electricity;
		• A floating version of the earlier on-shore Limpet concept, Figure 1.4. Project's first phase 1 MW floating grid-connected demonstrator planned for EMEC's Billia Croo, Orkney test site.

Appendix G

A tidal poem

Taken from Grace Helen Mowat's 1928 book *Funny Fables of Fundy*, with kind permission from the St Andrew's Civic Trust, New Brunswick, Canada:

> *A stranger once said to the tides in the Bay:*
> *'How strange you should live in this indolent way;*
> *You crawl up the strand then crawl down again*
> *Why can't you be useful and helpful to men?*
> *For the past thousand years you have been just the same,*
> *Such an idle existence! It's really a shame!'*
>
> *The tides, rather ruffled, cried "What do you wish?*
> *We fill up the fish weirs and bring in the fish*
> *And drift-wood and rock-weed and much else besides.*
> *Why, everyone waits for the turn of the tides!*
> *We've washed the shores clean and never once shirked*
> *If you did half as much you would feel overworked!"*
>
> *'I propose,' said the stranger (ignoring their theme),*
> *'To use all your strength in a practical scheme.*
> *I studied at college before I came here,*
> *And everyone thinks me a great engineer!*
> *I can hardly expect you to know who I am,*
> *But I'm seriously thinking of building a dam*
> *To keep you in bounds, till I need you, of course,*
> *And then I expect to control you by force.*
> *You can turn wheels and cranks by this simple device*
> *And greatly aid commerce. Now won't that be nice?*
>
> *'The waves made no answer to what the man said;*
> *But talking it over that evening in bed*
> *They grumbled and murmured: "We need not fear him;*
> *Beside our great strength his adventure looks slim.*
> *If he built up this, it is perfectly plain,*
> *We must all push together and break it again.*
> *And, if this arrangement should fail to survive,*
> *We can wash in a shark that will eat him alive!"*

The engineer tactfully waited awhile
Then, appearing next morning, he said with a smile:
'Dear tides, I am taking a trip up to town,
I hope you need something that I can bring down?'

They haughtily said: 'You may bring, if you wish,
Some good gelatine for the young jelly-fish.'

The item he added at once to his list,
And spoke of returning before he was missed;
And just as he promised, came home the next night,
His pockets all bulging with plans, blue and white,
The gelatine too he remembered to bring
(For jelly-fish need it so much in the spring!)
"These plans," he explained, "will be gold to your shore
By giving employment to men by the score."

But the tides in a voice that was hollow and cold,
Said: "Our fishes are silver; we don't care for gold."

'How hopelessly dull,' cried the great engineer.
'My college diploma is little use here!
I cannot express how this talk makes me feel!',
And appearing quite angry, he turned on his heel.

The sea-gulls brought word that a numerous band
Of workmen were filling the channel with sand,
And talked of erecting a barrier so high,
That no tides could cross over unless they could fly.

'Very well,' said the tides, "let him do as he will,
And we for a time will keep perfectly still
And wait for the Equinox gales in the Fall –
And they you will see what becomes of this wall!"

The sea-gulls that Autumn all gathered in flocks,
To await the return of the fall Equinox.
They were fighting for seats with the plovers and crows,
When all of a sudden the Equinox rose!

With rushing and roaring the tides came apace –
And dealt the great structure a slap in the face!

The engineer, viewing the frantic attack,
Admonished the tides that they better keep back!
But they cried, 'We are holding our annual ball,
When the Equinox comes for a dance in the fall.'

Then the tides with fantastic grimaces upreared,
And the engineer groaned, 'It is just as I feared!'

Down, down, went the dam and the sea-wall besides,
And the engineer fell with the wreck of the tides.
And the waves washed his pockets as clean as could be
And carried his plan and his gold out to sea.

He may have survived, for I know he could swim,
But the tides never more have been bothered with him.

MORAL:

These facts tell us plainly to look on all sides
Before we are tempted to tamper with tides;
And when we are strangers, wherever we go,
There's always a side that we still do not know;
And if we too suddenly start to reform
Our plans and our gold may be lost in the storm!

References

Advanced Mechanics and Engineering Ltd (1992) *Reliability and Availability Assessments of Wave Energy Devices*, Report ETSU 1690. Harwell: Energy Technology Support Unit, AEA Technology.

Alcorn, R. and O'Sullivan, D. (eds.) (2014) *Electrical Design for Ocean Wave & Tidal Energy Systems* (IET Renewables Series) London: IET.

Arabian-Hoseynabadi, H., Oraee, H. and Tavner, P.J. (2010) 'Failure modes and effects analysis (FMEA) for wind turbines', *International Journal of Electrical Power & Energy Systems*, 32(7), pp. 817–824.

Barstow, S.F., Mork, G., Lonseth, L. *et al.* (2003) 'World waves: Fusion of data from many sources in a user-friendly software package for timely calculation of wave statistics in coastal waters', *Proceedings of the 13th ISOPE Conference*, Oahu, Hawaii.

Billinton, R. and Allan, R.N. (1992) *Reliability Evaluation of Engineering Systems: Concepts and Techniques*. 2nd edn. New York, NY: Springer.

Birolini, A. (2007) *Reliability Engineering, Theory & Practice*. New York, NY: Springer.

Boyle, G. (2004) *Renewable Energy, Power for a Sustainable Future*. Oxford: Oxford University Press.

Carroll, J., McDonald, A. and McMillan, D. (2014) 'Reliability comparison of wind turbines with DFIG and PMG drives', *IEEE Transactions on Energy Conversion*, 30, pp. 663–670.

Chung, H.S., Wang, H., Blaajberg, F. and Pecht, M. (eds.) (2015) *Reliability of Power Electronic Converter Systems* (IET Power & Energy Series) London: IET.

Coronado, C. and Fischer, K. (2015) *Condition Monitoring of Wind Turbines: State of the Art, Users Experience and Recommendations*. Bremerhaven: Fraunhofer IWES.

Crabtree, C.J., Zappalá, D. and Hogg, S.I. (2015) 'Wind energy: UK experiences and off-shore operational challenges', *Proceedings of the Institution of Mechanical Engineers Part A: Journal of Power and Energy*, 229(7), pp. 727–746.

Cruz, J. (eds.) (2010) *Ocean Wave Energy*. Berlin: Springer.

Delorm, T.M. (2014) *Tidal Stream Devices: Reliability Prediction Models During Their Conceptual & Development Phases*. PhD Thesis. Durham University, Durham.

Delorm, T.M., Lu, Y., Christou, A. and McCluskey, P. (2016) 'Comparisons of off-shore wind turbine reliability', *Journal of Risk and Reliability,* 230(3), pp. 251–264.

Delorm, T.M., Zappalá, D. and Tavner, P.J. (2011) 'Tidal stream device reliability comparison models', *Proceedings of the Institution of Mechanical Engineers, Part O: Journal of Risk and Reliability*, 226(1), pp. 6–17.

European Marine Energy Centre, EMEC (2016). Available from: http://www.emec.org.uk/ (Accessed: June 2016).

Falcao, A.F. de O. (2010) 'Wave energy utilization: A review of the technologies', *Renewable and Sustainable Energy Reviews*, 14, pp. 899–918.

Faraci, V. (2006) 'Calculating failure rates of series/parallel networks', *Journal of the System Reliability Center*, First Quarter, pp. 1–7.

Faulstich, S., Hahn, B. and Tavner, P.J. (2011) 'Wind turbine downtime and its importance for off-shore deployment', *Wind Energy*, 14(3), pp. 327–337.

Feng, Y., Tavner, P.J. and Long, H. (2010) 'Early experiences with UK round 1 off-shore wind farms', *Proceedings of the Institution of Civil Engineers, Energy*, 163(EN4), pp. 167–181.

FINO Datenbank (2014) *Bundesamt fur Seeschifffahrt un Hydrographi (BSH).* Available from: https://www.bsh.de/EN/TOPICS/Offshore/offshore_node.html;jsessionid=EB27ACAC7FFBB4913EDDEC618990DA0F.live11293.

Fitton, G. (2013) Multifractal analysis and simulation of wind energy fluctuations, Thèse Doctorale de l'Université Paris-Est. Available from: https://theses.hal.science/tel-00962318/.

Flinn, J. and Bittencourt, C. (2008) 'Reliability estimation method for wave & tidal energy converters', *Proceedings of the 2nd International Conference on Ocean Energy (ICOE)*, Brest, France, 15–17 October.

Fusco, F. and Ringwood, J.A. (2009) 'Study on short-term sea profile prediction for wave energy applications', *8th EWTEC*, Uppsala, Sweden, September.

Fusco, F. and Ringwood, J.V. (2010) 'Short-term wave forecasting with AR models in real-time optimal control of wave energy converters', *IEEE Transactions on Sustainable Energy*, 1(2), pp. 99–106.

Goldberg, H. (1981) *Extending the Limits of Reliability Theory.* New York: Wiley.

Hahn, B., Durstewitz, M. and Rohrig, K. (2007) 'Reliability of wind turbines', *Proceedings of the Euromech Colloquium*, Oldenburg/Berlin: Springer, pp. 329–332.

Hardisty, J. (2009) *The Analysis of Tidal Stream Power.* New York: Wiley, 342 pages.

Hasselmann, K., Barnett, T.P., Bouws, E. *et al.* (1973) 'Measurements of wind-wave growth and swell decay during the Joint North Sea Wave Project (JONSWAP)', *Erganzungshaft zur Deutschen Hydrographischen Zeitschrift, Reihe A,* 8(12), pp. 1–95.

Hogg, S. and Crabtree, C.J. (eds.) (2017) *UK Wind Energy Technologies.* London: Routledge, Taylor & Francis Group.

IEEE, Gold Book (1997) *Recommended Practice for Design of Reliable Industrial and Commercial Power Systems.* Piscataway, NJ: IEEE Press.

International Energy Agency (2016) 'IEA wind task 33 reliability data'. Available from: https://iea-wind.org/task-directory/.

Krogstad, H.E. and Barstow, S.F. (1999) 'Satellite wave measurements for coastal engineering applications', *Coastal Engineering*, 37(3–4), pp. 283–307.

MacKay, D.J.C. (2008) *Under-Estimation of the UK Tidal Resource.* Cavendish Laboratory, University of Cambridge, Cambridge.

Mackie, G. (2008) 'Development of Evopod tidal stream turbine', *Proceedings of the International Conference on Marine Renewable Energy*, Royal Institute of Naval Architects, London, 9–17 November.

Meseguer, D., Yebra, N., Rasmussen, S., Weinell, C. and Thorslund Pedersen, L. (2010) 'Marine fouling and corrosion protection for off-shore ocean energy setups', *3rd International Conference On Ocean Energy*, Bilbao, 6 October.

MIL-HDBK-189 (1981) *Military Handbook: Reliability Growth Management.* Washington, DC: US Department of Defense.

MIL-HDBK-217F (1991) *Military Handbook: Reliability Prediction of Electronic Equipment.* Washington, DC: US Department of Defense.

MIL-HDBK-338 (1998) *Military Handbook: Electronic Reliability Design Handbook.* Washington, DC: US Department of Defense.

MIL-STD-1629A (1980) *Procedures for Performing a Failure Mode, Effects and Criticality Analysis.* Washington, DC: US Department of Defense.

Milan, P., Wachter, M. and Peinke, J. (2014) 'Stochastic modeling and performance monitoring of wind farm power production', *Journal of Renewable and Sustainable Energy*, 6, p. 033119.

Modarres, M., Kaminskiy, M.P. and Krivtsov, V. (2010) *Reliability Engineering and Risk Analysis: A Practical Guide.* 2nd edn. Boca Raton, FL: CRC Press.

Moir, I. (1998) 'The all-electric aircraft-major challenges', IEEE Colloquium on All Electric Aircraft. Digest No. 1998/260:2/1-2/6.

Moørk, G., Barstow, S., Kabuth, A. and Pontes, M.T. (2010) 'Assessing the global wave energy potential', *Proceedings of the OMAE2010 29th International Conference on Ocean, Off-Shore Mechanics and Arctic Engineering*, Shanghai, China.

Mukora, A.E. (2013) *Learning Curves and Engineering Assessment of Emerging Energy Technologies: On-Shore Wind.* PhD Thesis. Edinburgh University, Edinburgh.

Myers, L.E., Bahaj, A.S., Retzler, C. *et al.* (2011) 'Device classification template. Equitable testing and evaluation of marine energy extraction devices in terms of performance, cost and environmental impact, Deliverable D5.2, EquiMar', Commission of the European Communities, Brussels.

National Renewable Energy Centre, NAREC (2016) Available from: https://ore.catapult.org.uk/ (Accessed: June 2016).

Neij, L. (1999) 'Cost dynamics of wind power', *Energy*, 24(5), pp. 375–389.

NPRD-95 (1995) *Nonelectronic Parts Reliability Data.* Utica, NY: Reliability Information Analysis Center, RAIC.

O'Connor, M. and Dalton, D.J. (2012) 'Operational expenditure costs for wave energy projects and impacts on financial return', *Renewable Energy*, 50, pp. 1119–1131.

Off-shore & Onshore Reliability Data by OREDA Veracity (2024) *Marine Technology Consultants: SINTEF Industrial Management* Det Norske Veritas.

Okorie, O.P. (2011) *Scale Effects in Testing of a Monopile Support Structure Submerged in Tidal*. PhD Thesis. Robert Gordon University, Aberdeen.

OREDA (1984–2015) Dean, R.G. and Houston, J.R. (2013) 'Recent sea level trends and accelerations: Comparison of tide gauge and satellite results', *Coastal Engineering*, 75, pp. 4–9.

Pierson, W.J. and Moskowitz, L. (1964) 'A proposed spectral form for fully developed wind seas based on the similarity theory of S. A. Kitaigorodskii', *Journal of Geophysical Research*, 69(24), pp. 5181–5190.

Reliability Information Analysis Center (2010) *Reliability Modeling. The RIAC Guide to Reliability Prediction Assessment and Estimation*. Washington, DC: US Department of Defense.

Reliability Information Analysis Center and Data & Analysis Center for Software (2005) *System Reliability Toolkit: A Practical Guide for Understanding & Implementing a Program for System Reliability*. Washington, DC: US Department of Defense.

Renewables UK (2013) 'Wave & tidal energy in the UK, conquering challenges, generating growth'. Available from: www.RenewableUK.com (Accessed: February 2016.

Schmid, J. and Klein, H.P. (1991) *Performance of European Wind Turbines* (Applied Science) London & New York: Elsevier.

Sørensen, J (2014) Reliability assessment of wind turbines. In: R.D.J.M. Steenbergen, P.H.A.J.M. van Gelder, S. Miraglia, and A.C.W.M. Vrouwenvelder (eds), *Safety, Reliability and Risk Analysis: Beyond the Horizon* (pp. 27–36). London: CRC Press.

Spinato, F. (2008) *The Reliability of Wind Turbines*. PhD Thesis. Durham University, Durham.

Spinato, F., Tavner, P.J., van Bussel, G.J.W. and Koutoulakos, E. (2009) 'Reliability of wind turbine sub-assemblies', *IET Proceedings of the Renewable Power Generation*, 3(4), pp. 1–15.

Stiesdal, H. and Madsen, P.H. (2005) 'Design for reliability', *European Off-Shore Wind Conference*, Copenhagen.

Strategic Initiative for Ocean Energy (2014) 'Wave & tidal energy market deployment strategy for Europe'. *Co-Funded by the Intelligent Energy Europe, Programme of the European Union.*

Tavner, P.J. (2012) *Off-Shore Wind Turbines-Reliability, Availability & Maintenance* (IET Renewables Series) London: IET.

Tavner, P.J., Faulstich, S. and van Bussel, G.J.W. (2010) 'Reliability and availability of wind turbine electrical and electronic components', *EPE and Drives Association Journal*, 20(4), pp. 44–50.

Tavner, P.J., Ran, L., Penman, J. and Sedding H. (2008) *Condition Monitoring of Rotating Electrical Machines* (IET Energy Series) London: IET.

Tavner, P.J., Rigdon S.E. and Basu A.P. (2000) *Statistical Methods for the Reliability of Repairable Systems*. New York: Wiley.

Thies, P.R. (2012) *Advancing Reliability Information for Wave Energy Converters*. PhD Thesis. Exeter University, Exeter.

Thies, P.R., Flinn, J. and Smith, G.H. (2009) 'Is it a showstopper? Reliability assessment and criticality analysis for wave energy converters', *Proceedings of the 8th European Wave & Tidal Energy Conference*, Uppsala, Sweden, 7–10 October, pp. 21–30.

US Department of Energy (2009) 'Marine and hydrokinetic technology database'. Available from: http://energy.gov/eere/office-energy-efficiency-renewable-energy (Accessed: June 2016).

Val, D.V. (2009) 'Aspects of reliability assessment of tidal stream turbines', *Proceedings of the 10th International Conference on ICOSSAR*, Osaka, Japan, 13–17 September.

van der Hoeven, I. (1957) 'Power spectrum of horizontal wind speed in the frequency range from 0.0007 to 900 cycles per hour', *Journal of Atmospheric Sciences*, 14(2), pp. 160–164.

VGB PowerTech (2007) 'Guideline, reference designation system for power plants (RDS-PP); application explanation for wind power plants guideline, reference designation system for power plants (RDS-PP)', VGB-B 116 D2, VGB PowerTech, Essen, Germany.

WAMDI Group (1988) 'The WAM model – a third generation ocean wave prediction model', *Journal of Physical Oceanography*, 18, pp. 1775–1810.

Wikipedia (2016a) 'Accelerated life testing'. Available from: https://en.wikipedia.org/wiki/Accelerated_life_testing (Accessed: June 2016).

Wikipedia (2016b) 'Beaufort scale'. Available from: https://en.wikipedia.org/wiki/Beaufort_scale (Accessed: January 2016).

Wikipedia (2016c) 'EU ocean energy developments'. Available from: http://www.si-ocean.eu/en/upload/docs/SIOcean_Market_Deployment_Strategy-Web.pdf (Accessed: June 2016).

Wikipedia (2016d) 'Material fatigue'. Available from https://en.wikipedia.org/wiki/Fatigue_%28material%29#Miner.27s_rule (Accessed: January 2016).

Wikipedia (2016e) 'Wave power'. Available from: https://en.wikipedia.org/wiki/Wave_power (Accessed: January 2016).

Windstats (WSD & WSDK) (2010) 'Quarterly newsletter, part of Wind Power Weekly, Denmark'. Available from: www.windstats.com (Accessed: February 2010).

Weapons Center SD-18 (2006) *Program Guide for Parts Requirements and Application, Defense Standardization Program*. Crain, IN: Naval Surface Weapons Center.

Wolfram, J. (2006) 'On assessing the reliability and availability of marine energy converters: The problems of a new technology', *Proceedings of the Institution of Mechanical Engineers, Part O: Journal of Risk and Reliability*, 220(1), pp. 55–68.

Wood, J.K., Bahaj, A.S., Turnock, S.R., Wang L. and Evans, M. (2010) 'Tribolo-gical design constraints of marine renewable energy systems', *Philosophical Transactions of the Royal Society A*, 368, pp. 4807–4827.

Yang, W., Tavner, P.J. and Court, R. (2013) 'An online technique for condition monitoring the induction generators used in wind and marine turbines', *Journal of Mechanical Systems and Signal Processing,* 38, pp. 103–112.

Yang, W., Tavner, P.J., Crabtree, C.J., Feng, Y. and Qiu, Y. (2014) 'Wind turbine condition monitoring: Technical and commercial challenges', *Wind Energy*, 17, pp. 673–693.

YARD (1980) 'Reliability study of wave power devices'. Memorandum 3551/80, Report ETSU 1581. Harwell: Energy Technology Support Unit, AEA Technology.

Index

www.ingramcontent.com/pod-product-compliance
Lightning Source LLC
Chambersburg PA
CBHW050510190326
41458CB00005B/1486